# THE OPTICAL PERFORMANCE OF THE LIGHT MICROSCOPE

# List of titles

1. **A Short History of the Electron Microscope**
2. **Modern Electron Microscopes** (SEM, TEM: design, applications, limitations)
3. **Accessories for the Transmission Electron Microscope** (stages, apertures, cameras, image enhancement SAD, SSD, etc.)
4. **Preparation of Samples and Other Techniques for the Transmission Electron Microscope** (sectioning, staining, replication, etc.)
5. **Scanning Electron Microscopy** (sample prep, use of SEM, SSD, etc.)
6. **Specialized Electron Microscopes** (emission, reflection, high and low voltage)
7. **Field-Ion Emission Microscopes** (Mueller's work, one-atom probe)
8. **X-Ray Microscopy** (projection, microradiography, Kirkpatrick)
9. **Microprobes** (instruments, electron, ion, mini; design, maintenance, operation)
10. **Microprobes** (specimen preparation, techniques, automation)
11. **A Short History of the English Microscopes** (mechanical design emphasized)
12. **A Short History of American Microscopes** (mechanical design emphasized)
13. **A Short History of Light Microscopy** (techniques, top lighting, polarized light, dark field, thermal analysis, apochromats, fluorescence, interference, phase, dispersion staining)
14. **Performance of the Light Microscope, Part I**
15. **Performance of the Light Microscope, Part II**
16. **Accessories for the Light Microscope** (mechanical stages, micromanipulators, Lieberkuhn, micropolychromar, dispersion staining, demonstration ocular, hot stages, cold stages drawing cameras, reticles, fiber optics imagery, DTA, stereoscopy, microprojection)
17. **Special Methods in Light Microscopy** (increase resolving power, increase specimen contrast, sample characterization, specimen preparation, microscopy as adjunct to other techniques)
18. **Photomicrography** (stereo, Cine, serial section Cine)
19. **Photomacrography**
20. **Polarized Light Microscopy** (transmission)
21. **Polarized Light Microscopy** (reflection)
22. **Metallographic Techniques**
23. **Crystal Morphology**
24. **Microscopy in the Ultraviolet**
25. **Microscopy in the Infrared**
26. **Microspectrophotometry** (absorption and emission)
27. **Holographic Microscopy**
28. **Phase Microscopy**
29. **Interference Microscopy**
30. **Fluorescence Microscopy**
31. **Microtomy**
32. **Sections of Hard Materials** (thin and polished)
33. **Clinical Microscopy**
34. **Microphotography**
35. **Dispersion Staining**
36. **Thermal Microscopy 1-component system**
37. **Thermal Microscopy 2-component system**
38. **Micrometry**
39. **Stereology**
40. **Automatic Image Analysis**
41. **Mineral Identification** (thin sections)
42. **Mineral Identification** (grains)
43. **Microchemical Tests**
44. **Characterization of Single Small Particles**
45. **Study of Fibers**
46. **Study of Surfaces**
47. **Resinography**
48. **Liquid Crystals**
49. **Universal Stage**
50. **Integration of Microscopy into the Research Laboratory**
51. **Dictionary for Microscopy**
52. **Teaching Microscopy**

# THE OPTICAL PERFORMANCE
# OF THE LIGHT MICROSCOPE

Physical Optical Aspects
of Image Formation

H. Wolfgang Zieler
AtIantex and Zieler Instrument Corporation
Dedham, Massachusetts

Microscope Publications Ltd.
London, England
Chicago, Illinois

1974

Microscope Publications Ltd.

2 McCrone Mews, Belsize Lane
London NW3 5BG, G.B.

or

2820 South Michigan Avenue
Chicago, IL 60616 U.S.A.

Library of Congress Catalog Card Number: 72-85238

ISBN No. 0 904962 02 4

Printed by Newnorth Artwork Ltd. Bedford, England.

Dedicated to the memory of Professor Max Berek, a great ex-
perimentalist and teacher of optical crystallography early in
this century.




— about the author


## H.WOLFGANG ZIELER



Wolfgang Zieler died suddenly of a heart attack on Friday, 9th
February, 1973 at the age of 75 while on vacation in Puerto Rico.

He is survived by his wife, Loretta and two daughters, Nancy
and Ellen. A truly remarkable microscopist and man, he will
be missed by all who knew him or who had occasion to hear his
lectures or read his publications. His tremendous vitality and
enthusiasm belied his years but he was born in 1897 in Berlin
and had completed more than 50 professional years in Microscopy. Starting with Leitz in Wetzlar, Germany in January 1922
he spent a year gaining a background in microscope optics, design and construction. One of his teachers was Prof. M. Berek.

He travelled extensively throughout Europe and the near East for

LEITZ before coming to the U.S.A. in 1924. He then worked for
LEITZ in New York for 21 years, first as Manager of the Tech-
nical Department, then as President. In 1945, he moved to
Chicago and spent 10 years with W. H. Kessel Company as a
salesman of microscopes and educator of microscopists. In
1956, he started his own instrument sales company in Boston and
merged his company with the Atlantex Instrument Company in
1968. He was fully active as a working vice-president of Atlantex
and Zieler Instrument Company until his death.

During all of his career he was active as a teacher, lecturer and
author. He was active in the microscopical societies and held
various high offices in all of them. He was a Fellow, Life Member and Past President of the New York Microscopical Society,
Fellow of both the Royal Microscopical Society and the Biological
Photographic Association. He was a Member of the Optical Society of America, The State Microscopical Society of Illinois and
a Charter Member of the Electron Microscopical Society of
America. He has published more than 20 papers in microscopical,
optical and photographic journals and textbooks related to microscopy, photomicrography, microphotography and optics. His second volume on "The Optical Performance of the Light Microscope"
is his second book.

He was a frequent lecturer before scientific societies and
taught a number of courses in microscopy and photomicrography.
Some institutions which profited from his outstanding lecturing
ability and sound microscopical background included Brooklyn
Polytech, NYMS, SMSI, Northeastern University, New York University, New York University, City College of New York, University of Illinois, Notre Dame and MIT. He attended and presented
papers at most of the INTER/MICRO conferences from 1948
through 1972. He was several times banquet speaker at these
meetings and filled the same function or that of Chairman or
Master of Ceremonies at numerous other meetings.

He was first of all a professional microscopist; second, a bril-
liant teacher; third, a popular lecturer and writer; fourth, a
salesman of microscopy and finally, an instrument salesman.
Although he earned his living as an instrument salesman he

never "sold" instruments.  Instead, he helped his customers
solve problems microscopically.  If the equipment you needed
to solve a problem wasn't available he would modify, adapt or
design a unit that would do the job.  He would spend any amount
of time helping anyone who needed help.  He was obviously most
deeply concerned that you knew how and why an instrument worked and how to get the best out of it.

His zest for life itself as well as for anything microscopical was
infectious.  Time passed swiftly when Wolfgang Zieler lectured.
Buoyed by his own enthusiasm and the response of his audiences
he seldom heard the Chairman's time warnings or the end-of-
class-period bell — nor did we.

— reprinted from the Microscope, April 1973

H.WOLFGANG  ZIELER

1897 - 1973

# PREFACE

For years Wolfgang Zieler has had plans and notes for a book on microscope optics. During this period he has continuously refined his presentation of the necessary material by practicing on numerous students, lecture audiences and purchasers of microscopes. We are pleased to have catalyzed the production of two volumes on the optical performance of the light microscope based on Wolfgang's fifty years of experience in microscopy. The first volume on the geometrical optical aspects of image formation, Volume 14 in this series, appeared last June. Volume 15 now covers the physical optics of image formation and brings us to the point of achieving the ultimate in performance of the light microscope.

His concise and logical presentation will be useful to microscopists on all levels. It is a useful textbook for the self-educator and will be helpful to every teacher of microscopy.

Walter C. McCrone
Editor —— Microscope Series

Diffraction pattern of a nearly hexagonal array of dots (Abbe test plate) photographed with white light in the objective back focal plane.  A phase microscope annulus lies in the substage condenser lower focal plane.

VOLUME 15
## THE OPTICAL PERFORMANCE OF THE LIGHT MICROSCOPE

### PHYSICAL OPTICAL ASPECTS OF IMAGE FORMATION

CHAPTER 1
## THE PHYSICAL NATURE OF LIGHT

A. <u>INTRODUCTION</u>: In the companion volume to this (Volume 14) I endeavored to cover the geometrical aspects of image formation. The topics covered there included refraction and reflection, image formation, illumination, aberrations and the characteristics of oculars and objectives. Knowledge of these matters enables the microscopist to make the best use of his instrument.

In this volume I hope to explain the physical optics needed to understand how the microscope delivers the best final image. I will cover the nature of light, the distinction between coherent and incoherent light and between self-luminous and nonself-luminous objects, Abbe's diffraction theory, useful magnification and selection of equipment.

The science of physical optics explains phenomena arising under a variety of optical conditions, <u>e.g.</u>, the passage of light through one or more narrow slits, gratings or circular apertures of small diameter and even through optical instruments like the telescope and the microscope. Theories have been advanced which interpret the propagation of light as a wave motion. From time to time these theories have been modified when experiments aimed at confirmation or rejection of some of their details have revealed phenomena which made such revisions necessary.

The practicing microscopist need not study all of these historical developments and physical details of these theories. He should, however, gain knowledge of those aspects which serve for satisfactory explanation of the phenomena encountered in microscopy and how they influence the capacity of the microscope to reproduce object detail in the image — THE RESOLVING POWER — as well as the optical character of the image, formed under various optical conditions.

A helpful approach to the study of the physical optical aspects of image formation can be made by analyzing phenomena, produced by other types of wave motions; for instance, the waves spreading from a center of origin on the surface of water. Experiments can be made with water waves to disclose both analogies and differences between water waves and light waves.

B.  TRANSVERSE WAVE MOTION,  WAVELENGTH AND AMPLITUDE: When a stone drops into water, the equilibrium of the plane  smooth surface of this elastic "medium" is disturbed.  The stone, so to speak, pushes the water surface downward at the point of entry.  Because of the elasticity of the water, a vibratory wave motion starts.  To reestablish equilibrium, the water surface moves upward at that point, but passes beyond the original plane to a point above it, from which it moves downward again, swinging up and down, gradually approaching the "status quo", the smooth plane surface.

Adjacent points on the surface, at equal and increasing distances from the point where the stone has dropped into the water, follow the vibratory motion with slight and increasing time delays, thus creating a wave motion which spreads in all directions with equal speed.  It is visible as concentric circles of increasing diameters.

Whereas the <u>waves</u> spread in all directions, <u>away from</u> the center of origin, each <u>point</u> of the water surface <u>moves only up and down</u> with decreasing amplitude, in a single plane, perpendicular to the plane, undisturbed water surface and also to the direction of propagation of the waves.  Because of these characteristics, water waves are called TRANSVERSE WAVES.

Each circle of an expanding wave represents a disturbance of the water surface of equal magnitude and the expanding radii of the circles indicate the distances which the waves have travelled within a given time interval.  Each circle of a wave is called the WAVEFRONT.  The constant distance from crest to crest of adjacent wavefronts is called the WAVELENGTH.  It is generally measured from a point on the undisturbed water surface, over the crest of the wave, back to the same level — this part representing positive phase — and through the trough and again back to the original level — this part being the negative phase of the wave.

The amplitude is the distance between the original water level and the wave crest.  As the wave motion expands in all directions, the amplitude decreases rapidly and uniformly along the expanding wavefronts.  This is shown in a diagrammatic "top view" of the water surface a short time after the wave motion started (Figure 1).

The circular fronts of the crests are shown as drawn-out lines, the troughs as dotted lines.  The lower part of Figure 1, a side view of a wave, shows the wavelength $\lambda$ and the amplitude $\alpha$.

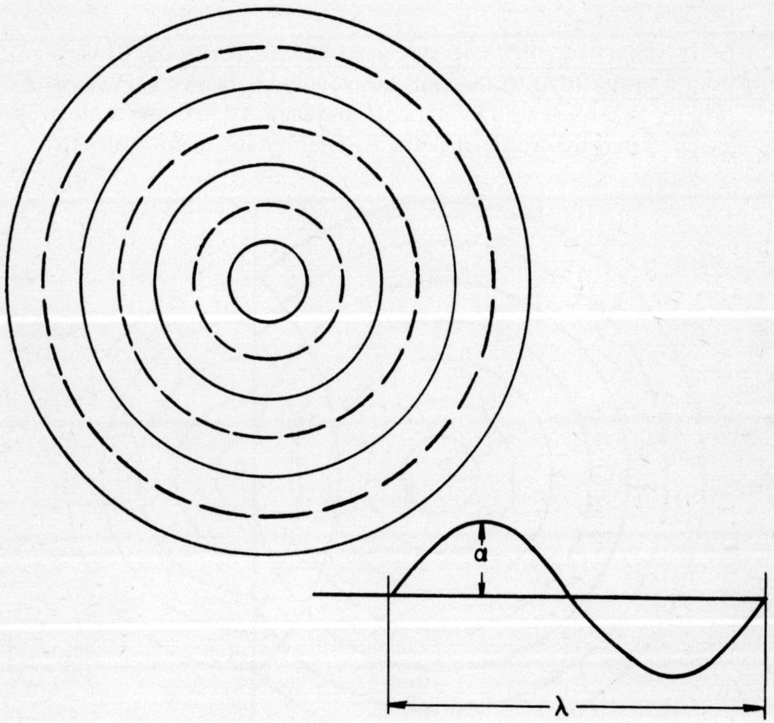


Figure 1.  Water waves spreading from origin at center.  At lower right the wavelength ($\lambda$) and amplitude ($\alpha$) are shown.

C.  PASSAGE OF TRANSVERSE WAVE MOTION THROUGH A
NARROW SLIT, DIFFRACTION: The following experiment
reveals an important physical property of water waves.  A
vertical wall is placed across a large pool filled with water.  In
this wall there is a narrow slit as point A (Figure 2).  A stone
is dropped into the water at point P, starting a wave motion.
The wall reflects these waves, except at point A where the
wave can pass through the wall.  This point now becomes the
center of a new wave motion proceeding in that space with the
same speed and wavelength, but with reduced amplitude.  This
phenomenon is called DIFFRACTION.  Each point of the original wavefront is also a potential origin for diffracted waves.
(Figure 3.)

D.  PASSAGE THROUGH TWO ADJACENT NARROW SLITS,
INTERFERENCE: If the wall is provided with two narrow slits
(Figure 4), separated by a small distance AB = D and a stone is
dropped into the water at point P, equidistant from A and B,

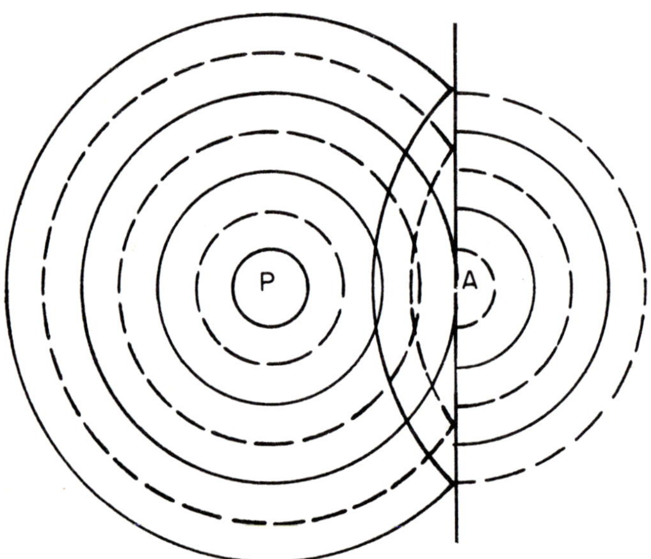


Figure 2.  Diffraction of transverse waves passing through
a single narrow slit.

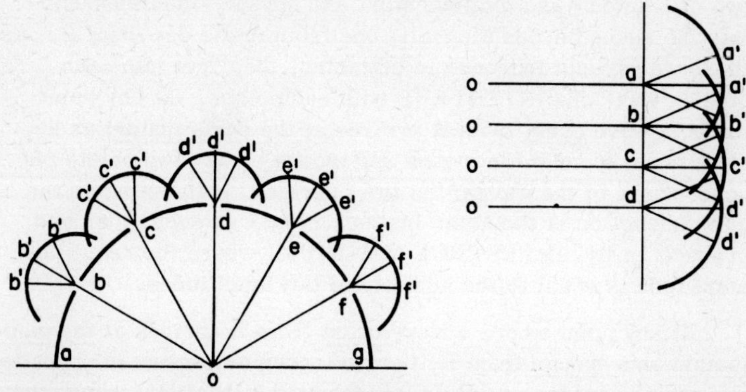

Figure 3. Potential centers of diffracted waves on a spherical wavefront (left) and a plane wavefront (right).

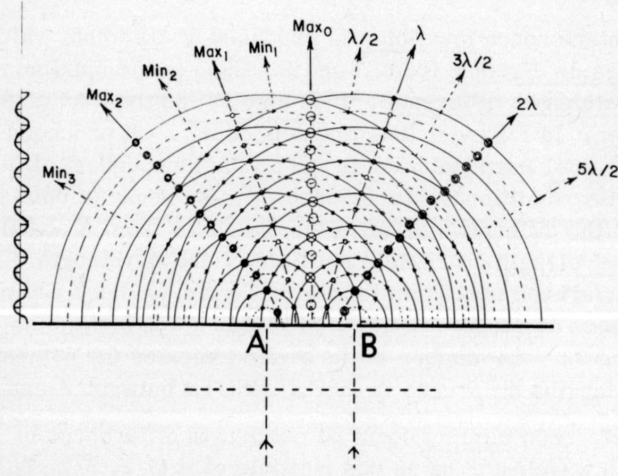


Figure 4. Interference of diffracted waves originating in two narrow slits.

two diffracted wave motions start and spread simultaneously
into the space beyond the wall, one from A and one from B.  As
these wavefronts increase in diameter, they "get into each
other's way" and INTERFERE with each other.  At any point
where a wave crest from A arrives at the same instant as a
wave crest from B (or trough and trough or any two points of
equal phase of the waves) the two energies, both acting in the
same direction at the same instant, reinforce each other and
produce an INTERFERENCE MAXIMUM, where the resulting
amplitude is equal to the sum of the two amplitudes.

At any point where a wave crest from A arrives at the same
instant as a trough from B, the two energies, acting in opposite
directions, produce an INTERFERENCE MINIMUM, where the
resulting amplitude is equal to the difference between the two
amplitudes.  When both amplitudes are of the same magnitude,
the result at the point of the interference is complete cancella-
tion of the two wave motions.

It has been assumed in this diagram that a stone has
dropped into the water at a point which is so far away from AB
that the small portion of the circular wave front arriving at the
wall is practically a straight line, perpendicular to the direction of the wave motion.

An interference maximum is produced at all points which
have the same distance from A and B.  This is the maximum
without pathlength difference.  It is also called the maximum of
zero order.  In Figure 4 it is marked "$Max_0$."  It proceeds in
a straight line, perpendicular to AB and is independent of the
wavelength.  Additional interference maxima occur at points
for which the difference between the distances from A and B is
one full wavelength or a whole multiple of one wavelength.  The
first interference maximum is at $\lambda$ in Figure 4.  Each additional difference of a full wavelength produces a maximum of a
higher order.  The number of the order indicates the number of
full wavelengths difference in PATHLENGTH between A and B.

Interference minima occur at pathlength differences of
one-half a wavelength or an odd multiple of it (3/2, 5/2, 7/2
etc.).  The maxima and minima proceed along hyperbolic paths
which depend on the distance between A and B as well as the

wavelength. At very long distances, compared to the wavelength, the directions are those of the asymptotes and can be interpreted as straight lines.

The geometrical relationship between the distance D from A to B, the wavelength $\lambda$ and the angle $\alpha$ between the maximum of zero order and the first maximum can be determined with the aid of Figure 5, in which the directions of the maxima are indicated by those of the asymptotes.

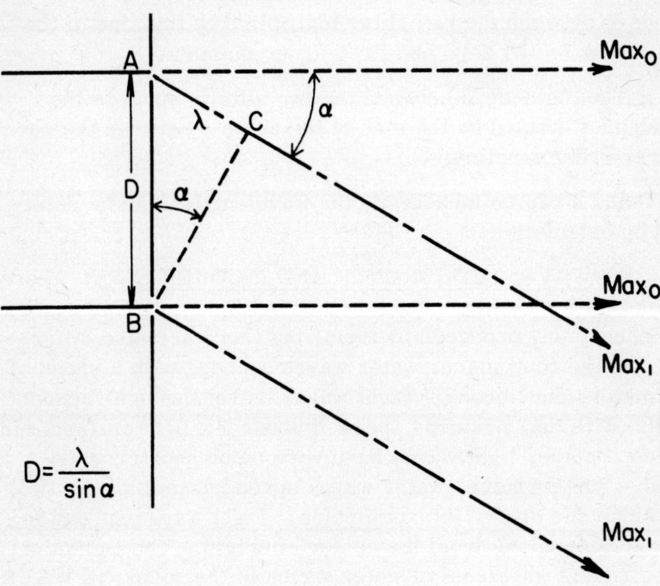


Figure 5. Geometrical relationship between directions of interference maxima and distance between two narrow slits.

The angle between $Max_0$ and $Max_1$ is $\alpha$. It is also equal to the angle ABC. The distance between A and B is D and the wavelength $\lambda$ is AC.

Therefore: $\sin \alpha = AC/AB = \lambda/D$

and from this: $\lambda = D \sin \alpha$ and $D = \lambda/\sin \alpha$          Equation (1-2)

These equations state that:

1. The sine of the angle $\alpha$ between the zero and first-order maxima is equal to the wavelength $\lambda$ of the wave motion, divided by the distance D between the two slits.

2. The wavelength $\lambda$ of the wave motion is equal to the distance D between the two slits, multiplied by the sine of the angle $\alpha$ between the zero and first-order maxima.

3. The distance D between the two slits is equal to the wavelength $\lambda$ divided by the sine of the angle $\alpha$ between the zero and first-order maxima.

Since D and $\alpha$ can be measured, the wavelength $\lambda$ can be determined by calculation.

E. <u>LIGHT WAVES, COHERENCE AND INCOHERENCE</u>: There are similarities between water waves and light waves in regard to the phenomena produced by them, but there are also differences. In the first place, water waves proceed with a speed of a few meters per second. Light waves travel through vacuum (and air) with the enormous speed of about $3 \times 10^{10}$ cm/sec. In the second place, light waves have very much shorter wavelengths. Furthermore, water waves spread essentially in two dimensions and are visible as <u>circles</u>. Light waves spread in all three dimensions from the center of origin. The equivalent of the <u>circular</u> wavefront of water waves is the <u>spherical</u> WAVE SURFACE of light waves. Light waves are called SPHERICAL WAVES.

The most significant difference between water waves and light waves, a difference of fundamental importance, can be demonstrated with another experiment. If, in the set-up for the experiment shown in Figure 4 <u>two</u> stones had been dropped into the water simultaneously, at two points equidistant, respectively, from A and B, the wave motion from the first

point would have arrived at A at the same instant the wave
motion from the second point arrived at B. Two sets of diffracted waves would then have spread simultaneously into the
space beyond the wall and would have produced exactly the same
pattern of interference maxima and minima as that shown in
Figure 4.

If, however, in an experiment with <u>light</u> waves, the slit at
A had been illuminated with light from one source and the slit
at B from a separate source, and if the distance from A to the
first light source was exactly equal to that from B to the second
light source, no interference maxima or minima would have
occurred. <u>Light waves can interfere with each other only when
they have one common origin</u>. This fact indicates that light
waves do not vibrate in a single azimuth, like water waves.
Although they vibrate in a plane, perpendicular to the direction
of propagation, the azimuth in which they vibrate within that
plane changes from one extremely short instant to another. In
Figure 6 it is assumed that the direction of propagation of the

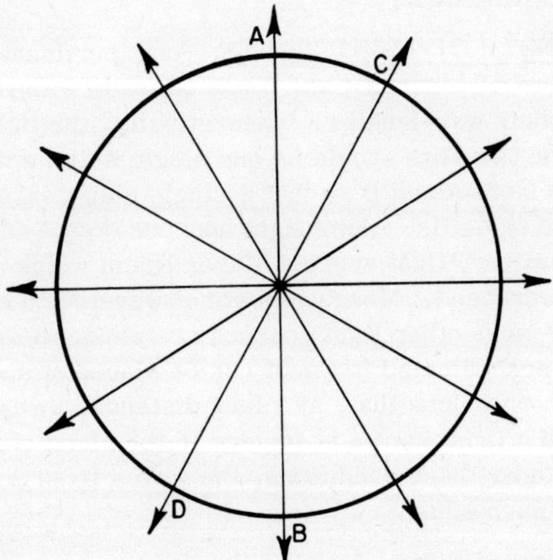

Figure 6. Changes of azimuths of vibration of
spherical waves.

light is perpendicular to the diagram. At one extremely short instant, the vibrations may occur in the azimuth AB. In the next instant, they may occur in the azimuth CD etc. The change of the azimuth of vibration is of such nature that it is repetitious with the wavelength. That means that light waves from one <u>common origin</u> and diffracted at two points of one and the same wave surface interfere with each other at any point where they arrive <u>simultaneously</u>, no matter how long the path-length difference between the two wave motions. When they originate from <u>two independent origins</u>, they do <u>not</u> interfere with each other, because at any given instant, they do not vibrate in the same azimuth.

This physical relationship between light waves originating from one common origin is called COHERENCE. When one single light source illuminates the two slits as shown in Figure 4, the two centers of the diffracted waves at A and B are points of the <u>same</u> "coherent" wave surface and traceable to one single center of origin. When there are two independent light sources illuminating the two slits, the physical condition at those points is called INCOHERENCE.

F. <u>WAVELENGTH OF LIGHT WAVES</u>: An experiment, based on the conditions of Figures 4 and 5 can be made with light waves to determine their wavelengths. Theoretically, the light source illuminating the two slits should be one single self-luminous point far away from A and B so that a single plane wave surface illuminates the two slits. Point light sources do not exist. There are, however, light sources (laser) from which plane wave surfaces proceed. Measurements of wavelength can, however, be made when other light sources are used. A small light source with a concentrated filament can be placed at such a distance from a convex lens that, at a long distance, a real magnified image of the light source is formed in the plane of the slits at A and B. Under these conditions, the waves from A and B interfere with each other.

The diagram of Figure 7 shows the experimental setup for measuring wavelengths. Two very narrow slits, parallel to each other and separated by the small distance D, are mounted on the front surface of a lightproof box. The rear wall of the box is at the long distance, t. A ground glass is placed in the

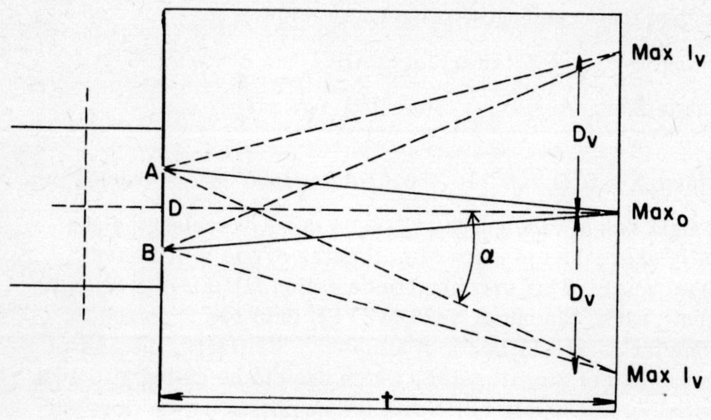


Figure 7. Determination of wavelength of light waves.

rear wall of the box. Light from a suitable source, emitting
white light, illuminates the two slits. An interference maximum of zero order, $Max_0$, is visible as a white elongated spot
on the ground glass. On both sides of the white center, the
light intensity diminishes and changes color until there is darkness where the first interference minimum is produced. At
equal distances to the right and left from the white band, the
maxima of the first order of the colors of the spectrum appear
with the color of the violet light closest to the central maximum.
This experiment reveals that the wavelength of light changes
with the spectral color.

When the distance D between the slits and the distance t
from front to rear wall are known, it is only necessary to meas-
ure the distance between the center of the maximum of zero
order and the first interference maximum of a selected color,
e.g., violet light, and the wavelength of that color can be calculated. Let the distance D between A and B be 0.01 mm and the
distance t be 1000 mm. The distance D between the center of
$Max_0$ and the interference maximum $Max_{1v}$ has been determined

by measurement to be 40 mm. The calculations, based on the conditions of Figure 5 for a distance t which is very much longer than the wavelength, proceed as follows:

According to Figure 5: in triangle ABC: $\sin \alpha = \lambda/D$

In triangle $\mathrm{Max}_0\text{-}A\text{-}\mathrm{Max}_{1v}$: $\sin \alpha = D_v / \sqrt{t^2 + D_v^2}$

Therefore: $\lambda = D.D_v / \sqrt{t^2 + D_v^2} = 0.01(40)/\sqrt{1000^2 + 40^2} = 0.0004$ mm

Violet light has a wavelength of 0.0004 mm. Wavelengths are generally expressed in units of millionths of one millimeter (0.000001 mm). This unit used to be called MILLIMICRON but, in recent years, the name NANOMETER (nm) has replaced it. A nanometer is one-billionth of a meter. Similarly, the unit of one thousandth of a millimeter, which used to be called MICRON is now called MICROMETER (with the emphasis on the first syllable). This is one-millionth of a meter. There is still another unit of length used frequently for wavelength. It is called ANGSTROM. One Angstrom is one-tenth of a nanometer or, one ten-billionth of a meter. It is hardly believable to the layman that the wavelength of a selected color of the spectrum can be determined with an accuracy of small fractions of an Angstrom. Incidentally, it is planned that the use of Angstrom will cease and that the nanometer will become the unit for small distances in the $10^{-9}$ meter range.

Light of one single wavelength is called MONOCHROMATIC. Wavelengths in the visible spectrum extend from about 400 to about 700 nm and the well known colors of the spectrum are within the following approximate ranges of wavelength:

|  |  |
|---|---|
| violet: | 400–455 nm |
| blue: | 455–492 nm |
| green: | 492–550 nm |
| yellow: | 550–588 nm |
| orange: | 588–647 nm |
| red: | 647–700 nm |

The human eye is not sensitive to wavelengths shorter than about 400 nm. These wavelengths, termed ULTRAVIOLET, can, however, be detected and measured by other means, _e.g._, photographic methods. The ultraviolet extends from the longest wavelength of about 400 to shorter than 200 nm. Glass transmits only a narrow range of the longest ultraviolet waves. Quartz, calcium fluoride and other materials transmit much shorter wavelengths. Even air absorbs ultraviolet waves shorter than about 200 nm.

The INFRARED spectrum, adjacent to the longest red wavelengths of the visible spectrum, covers a much wider range. Spectrophotometers are available for the infrared spectrum with dispersion prisms made of special crystalline materials like cesium bromide, potassium bromide, sodium chloride and calcium fluoride.

The refractive indices of transparent materials are generally defined for specific wavelengths rather than the name of the color. The refractive indices of two kinds of glass for specific wavelengths are listed in Table I.

Table I

Refractive indices of crown and flint glass

| Name of glass | Wavelength (nm) | | | | | |
|---|---|---|---|---|---|---|
| | 768.2 | 656.3 | 589.3 | 546.1 | 486.1 | 435.8 |
| Crown $BK_7$ | 1.51098 | 1.51385 | 1.51633 | 1.51894 | 1.52191 | 1.52629 |
| Flint $F_2$ | 1.60959 | 1.61504 | 1.62004 | 1.62410 | 1.63210 | 1.64206 |

Experiments for the determination of wavelength are generally carried out with linear GRATINGS rather than parallel slits. A linear grating can be interpreted as the result obtained by adding many more parallel slits, spaced at equal distances D from each other. Transmission gratings are generally produced by engraving fine parallel straight lines on a transparent material (glass, quartz etc.); reflection gratings on an opaque metal plate. The geometrical relationships between wavelength, distance between adjacent lines of the grating (the GRATING CONSTANT) and the angle between the maxima of zero and first order, are the same as those for the experiment with two slits, but the maxima produced by gratings are much narrower and the intensities are considerably higher.

G. EXPERIMENTS WITH LIGHT WAVES:

1. Passage of a coherent wave surface through a slit of

<u>finite width</u>:  In describing the preceding experiments, it was assumed that the slits were extremely narrow so that each could be interpreted as a single line of centers of diffracted waves.  If light passes through a single slit of <u>finite</u> width, for instance of the same width as the distance between two slits at A and B in Figure 5, a significant change in the pattern of interference maxima and minima occurs (Figure 8).

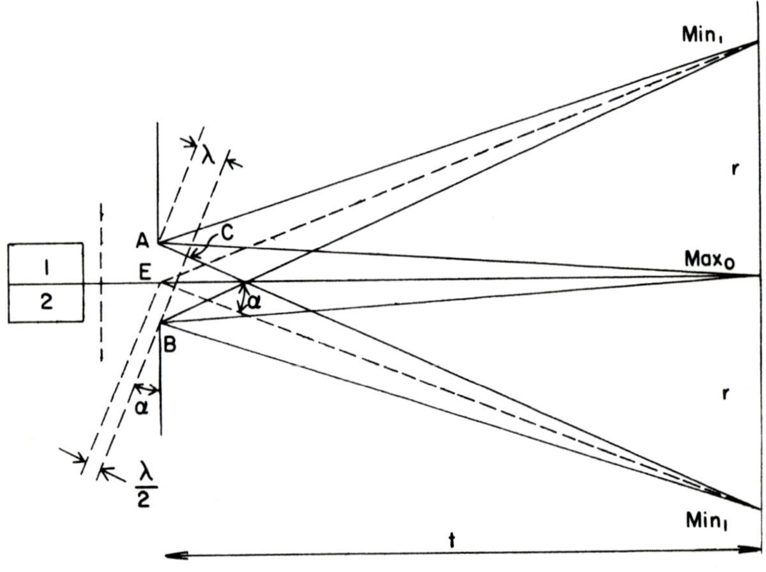


Figure 8.   Interference maximum $Max_0$ and first minimum $Min_1$ caused by passage of coherent plane wave surface through a single slit of finite width.

When the wave surface (the straight, dotted line in Figure 8) passes through the slit diffraction occurs, not only at points A and B, but also along the entire wave surface passing through the rectangular area of the slit.  On a screen at a distance, t, from the slit, an interference maximum of zero order occurs at point $Max_0$, equidistant from A and B, and from all of the corresponding points of the two halves of the wave surface between A and B.

The first interference <u>minimum</u>, however, occurs at a point for which the pathlength difference for A and B is a <u>full wavelength</u>. At points $Min_1$ the pathlength difference between $Min_1$-A and $Min_1$-E in the <u>center</u> between A and B is <u>one-half</u> wavelength. The pathlength difference for all points of the wave surface between A and E and all of the corresponding points of the other half between E and B is <u>one-half</u> wavelength. The <u>total</u> effect of all interferences with a half wavelength-pathlength difference which have passed through the entire width of the slit is the production of a minimum at $Min_1$ even though the total pathlength difference between $Min_1$ and the points A and B is <u>one full</u> wavelength.

At the left side of Figure 8 a cross-section through the rectangular area of the slit is shown, revealing two halves of <u>equal areas</u>. The significance of this equality will become evident during explanation of the next experiment.

2. <u>Passage of a coherent wave surface through a circular aperture</u>: In the experiments with slits, the maxima and minima lie along straight lines on both sides of the central maximum. When the wave surface passes through a <u>circular</u> aperture of finite diameter, the interference maxima and minima lie on concentric circular areas. However, the distance between $Max_0$ and $Min_1$ is not equal to that between $Max_0$ and $Min_1$ in the experiment with the rectangular slit but slightly longer.

This fact can be explained by making a comparison between four areas of equal width of the circular aperture, as shown on the left side of Figure 9. Waves diffracted from points throughout area No. 1 interfere with waves diffracted from points of area No. 3 of equal width, but greater total <u>area</u>. Waves diffracted from points of area No. 2 interfere with those from points of area No. 4 which has the same width but a smaller area. The total effect of all interference with half a wavelength-pathlength difference is <u>not</u> a minimum at a point for which the pathlength difference from diametrically opposite points of the aperture is one full wavelength because, so to speak, when waves diffracted from all of the points of one area have interfered with the corresponding waves from points of the corresponding area in the other half, there are still waves from points of the wave surfaces from the larger areas "left over".

Experiments, measurements and calculations have revealed that the first interference <u>minimum</u> occurs at a lateral distance r from $Max_0$ when the pathlength difference to points A and B of the circular aperture is 1.22 $\lambda$ (Figure 9).

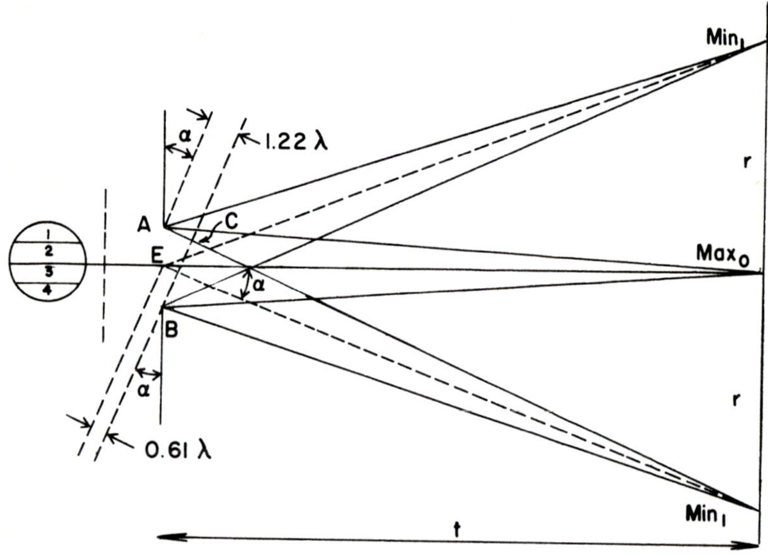


Figure 9.  Interference maximum $Max_0$ and first minimum $Min_1$ caused by passage of a coherent plane wave surface through a circular aperture of finite diameter.

The following geometrical relationships exist between the diameter of the circular aperture, D, the distance t to the screen, the wavelength $\lambda$ and the angle $\alpha$ between the directions from $Max_0$ and $Min_1$.

In triangle ABC:   $\sin \alpha = 1.22\lambda D$

In triangle $\text{Max}_0\text{-E-Min}_1$:   $\sin \alpha = r/\sqrt{t^2+r^2}$

(assuming that the screen is very far away, compared to the wavelength, the directions $A\text{-Max}_0$ and $B\text{-Max}_0$ are parallel to each other).

Therefore:   $1.22\lambda D = r/\sqrt{t^2+r^2}$

and:   $r = 1.22\lambda t/\sqrt{D^2-(1.22)^2\lambda^2}$

Since the factor, $(1.22)^2\lambda^2$, is extremely small, <1 nm, it can be neglected for practical purposes and the simplified equation expressing the geometrical relation between the factors involved is:

$$r = 1.22\lambda t/D \qquad\qquad \text{Equation (3)}$$

The pattern of light on the screen is a circular bright area in which the intensity decreases radially to a minimum at the radial distance r from the center of brightness.  The smaller the diameter of the circular aperture, the larger the diameter of the bright circular area.  That is the reason the image produced by the pinhole camera breaks down for very small "pinholes".

# IMAGE FORMATION OF SELF-LUMINOUS OBJECTS

A.  <u>INTRODUCTION</u>:  In the presentation of geometrical optical aspects of image formation (Volume 14), several temporary assumptions were made in the interest of the progressive explanation of problems involved.  Among these assumptions was one which regarded the object as consisting of an infinite number of <u>self-luminous</u> points.

The presentation of the <u>physical</u> optical aspects of image formation will start again with the analysis of the formation of images of self-luminous points and self-luminous objects of finite size.  This analysis, however, is of practical significance because there are objects which possess the optical property of self-luminosity.  These objects emit light of relatively long wavelengths; when they are illuminated with light of shorter wavelengths they fluoresce.  Furthermore, the light source itself, which is part of the optical system for illumination of nonself-luminous objects, is truly self-luminous.  In the passage of light through the illumination system, the object plane and the image-forming system, the wave surfaces originating from single points of the light source undergo modification and the optical conditions in the object plane can change from incoherence to coherence.  Nonself-luminous objects can assume the optical properties of self-luminosity when the illumination system is used for production of incoherence in the object plane.

These possible changes of the optical conditions in the object plane influence not only the capacity of the microscope to reproduce small object detail in the image (resolving power) but also the entire optical character of the image.

B.  <u>GEOMETRICAL AND OPTICAL PATHLENGTH</u>:  According to the law of refraction, the change of direction which occurs when light passes obliquely from a rarer into a denser medium is due to <u>reduction</u> of its <u>speed</u> in the denser medium.  Light waves of a given wavelength vibrate with a given <u>frequency</u> as they traverse the rarer medium.  This frequency does <u>not</u> change as light proceeds through the denser medium.  Therefore, the wavelength which was $\lambda$ in the medium with refractive index 1.00 changes to the <u>shorter</u> wavelength $\frac{\lambda}{n}$ as light proceeds through the medium with the refractive index n.  This is

shown in Figure 10 for passage of light from air (n = 1.00) to
water (n = 1.33).

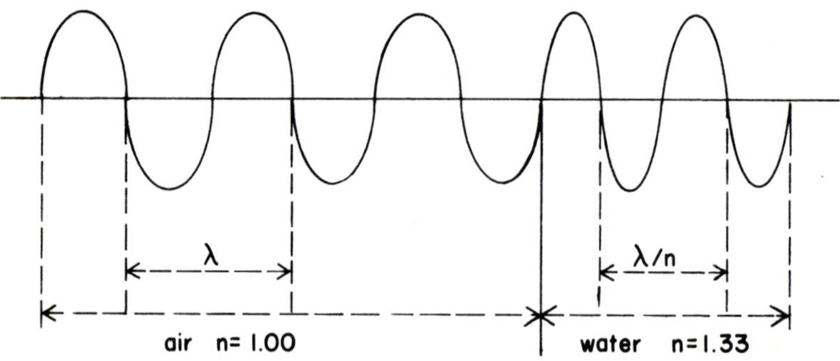


Figure 10.   Change of wavelength in passage of light from
              rarer into denser medium.

The light waves which have travelled a given distance, for
instance 100 mm, through air travel within the same time interval and with the same number of wavelengths through 100/1.33
= 75 mm of water (n = 1.33).

Although the GEOMETRICAL PATHLENGTH through air is
longer than that traversed within the same time interval through
water, the OPTICAL PATHLENGTH (the total number of wavelengths traversed within the same time interval) is equal in both
media.

C.  PASSAGE OF LIGHT FROM A SELF-LUMINOUS POINT
THROUGH A CONVEX LENS: In Figure 11, light from the self-
luminous point, A, on the optical axis proceeding along the
optical axis, enters the lens at B.  The spherical wave surface
is indicated by the dotted line which is part of a circle.  From
B to C, light travels through the denser medium with its speed
reduced from v in air to v/n in the glass of the lens.  Upon re-
entry into air, the speed is again v.

Another point of the same wave surface has proceeded at an
angle of inclination to the optical axis and has arrived at B'
when the first point has arrived at B.  From B', light proceeds
in the same direction and with the same speed through air until

it enters the lens at E.  Because of oblique incidence, light
is refracted at E and proceeds through the lens in a new direction to point F.  Because of the greatly reduced thickness of the
lens near its edge, the pathlength of the waves through the denser medium from E to F when the speed is reduced is shorter
than that from B to C.

At F, the light is again refracted.  It enters air and proceeds with its original speed.  Its path intersects the optical
axis at G.  At the same instant, the point of the wave surface
which had travelled from B to C has reached point D on the optical axis.

$$AB + nBC + CD = AE + nEF + FG$$

The relative lengths of the <u>retardations</u> of the points of the wave
surface passing through the lens from along the optical axis to
regions near its edge are such that the emerging wave surface
is not spherical and does not proceed to a single point of the
image space.  Spherical aberration occurs.  In Figure 11, the
<u>geometrical</u> pathlength from A via B and C to D is shorter than
that from A via B', E, F to G, but the <u>optical pathlengths</u> are
<u>equal</u>.

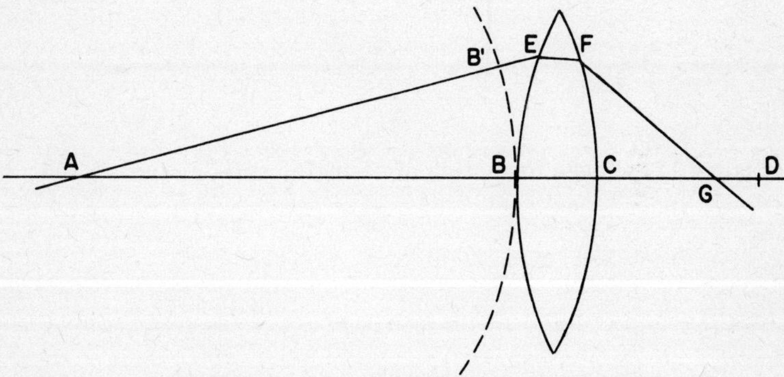

Figure 11.  Geometrical and optical pathlength in passage of
light through a convex lens.

D. <u>IMAGE FORMATION OF A SELF-LUMINOUS POINT</u>:  If a complex lens system, for instance, the objective of a microscope, is of theoretical perfection, the relative retardations of the points of a spherical wave surface (originating in a single, self-luminous point A on the optical axis) which occur in the passage through the lens system are of such varying magnitudes that, on emerging from it, the wave surface is again part of a sphere which proceeds toward its new center at the image point I.  At that point, all of the light, which was emitted at point A at a given instant and which was collected by the lens system, arrives simultaneously with equal optical pathlengths, producing an interference maximum of zero order (Figure 12).

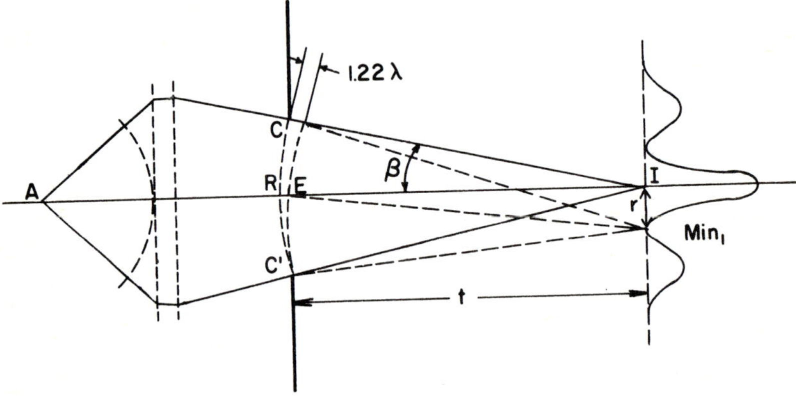


Figure 12.  Image formation of a single, self-luminous point.

The limited portion of the spherical wave surface, indicated by the dotted line, has emerged from the lens system and has passed through the aperture stop C - C' in its back focal plane. The center of the interference maximum at I is equidistant from C and C'.

In circular areas around I, at increasing distances in the image plane, the difference between the distances from C and C' increases. At point $Min_1$, at the distance r from I, the pathlength difference between $Min_1$ to C and to C' is $1.22\lambda$. All of the waves diffracted from the coherent portion of the wave surface between C and E interfere with those of the other half of the emerging wave surface from E to C' with such pathlength differences that they produce at $Min_1$ an interference <u>minimum</u> of the first order.

The image of point A which the lens has formed is a bright "disc" of light with the highest intensity at I, decreasing in all directions to a minimum at $Min_1$. This image is called a DIFFRACTION DISC. The geometrical relationships between the radius r of the diffraction disc (from I to $Min_1$), the radius R of the aperture stop (CE), the distance t from the center of the aperture stop to the center of the image (EI) and the angle, $\beta$ = CIE, (Figure 12) are:

$$r = 1.22\lambda \sqrt{R^2 + t^2}/2R$$

$$\sin \beta = R/\sqrt{R^2 + t^2}$$

Therefore:   $r = 1.22\lambda/2 \sin \beta$

and the diameter, d, of the diffraction disc is:

$$d = 1.22\lambda/\sin \beta \qquad \text{Equation (4)}$$

where $\lambda$ is the wavelength of the image-forming light and $\beta$ is the angle between the optical axis and the direction of the light having the greatest inclination still able to pass through the objective. This equation states that for image formation <u>at the distance t</u> from the plane of the aperture stop the <u>diameter</u> of the <u>diffraction disc decreases</u> as the <u>wavelength</u> of the image-forming light <u>decreases</u> and the angle $\beta$ increases; consequently, the <u>diameter of the aperture stop</u> also <u>increases</u>.[*]

---

[*]At correspondingly greater distances from point I (Figure 12) in a plane perpendicular to the optical axis, maxima and minima of higher orders occur but have much lower respective intensities.

E.  IMAGE FORMATION OF TWO OR MORE SELF-LUMINOUS
POINTS:  RESOLVING POWER:  When there are two (or more)
self-luminous points in the object plane, separated by a distance, $D_O$, their images, the two diffraction discs, will be separated by a distance, d, equal to their diameter.  The determination of these geometrical relationships is based on fulfillment
of the "sine condition", an essential requirement for image formation of objects of finite size to which reference was made on
page 70 of Volume 14.

When the wavelength of the light traversing the optical
medium of the refractive index n in the object space is reduced
from $\lambda$ to $\lambda/n$ in immersion systems equation (4) becomes:

$$D_O = 1.22\lambda/n \sin\beta \qquad\qquad \text{Equation (5)}$$

Since the numerical aperture of the objective is defined as
NA = n sin $\alpha$, (see Volume 14, pages 43–45) the equation is
changed once more to:

$$D_O = 1.22\lambda/NA \qquad\qquad \text{Equation (6)}$$

When the distance, $D_O$, between adjacent (self-luminous) <u>object</u>
points is equal to the value expressed in Equation (6), the centers of the diffraction discs in the image plane are separated by
a distance, equal to their diameters and the light intensity in
the center of the space between them is reduced to a <u>minimum</u>.
The variations of the intensity in the image plane are shown in
Figure 13.

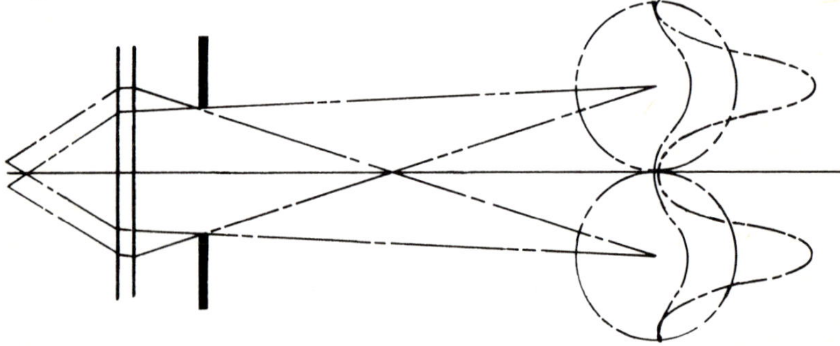

Figure 13.  Image formation of two self-luminous points.  The
           diffraction discs are separated by a distance equal
           to their diameters.

Under these conditions, the limit of the resolving power has not yet been reached. If the distance $D_0$ decreases still further, the two diffraction discs "overlap". In the space between the two maxima of the diffraction discs, there is an area in which both object points contribute to the light intensity. Since light from two independent centers of origin cannot interfere with each other, each of the object-points contributes to the total light intensity in that area. The total intensity is equal to the square of the amplitudes of the waves from the first point plus the square of the amplitudes of the waves from the other point. Even under these conditions, it is still possible to detect in the image that there are <u>two</u> points of origin in the object plane, as long as the intensity in the center between the maxima of intensity in the center between the maxima of intensity of the two diffraction discs is still sufficiently lower to be detectable by the observer.

Experiments, measurements and calculations have resulted in finding that even when the distance between the maxima of light intensity of adjacent diffraction discs is reduced to a magnitude, approximately equal to the <u>radius</u> of a disc so that the <u>maximum</u> $Max_0$ of one disc coincides with the first <u>minimum</u> of the adjacent disc (Figure 14), it is still possible to "resolve"

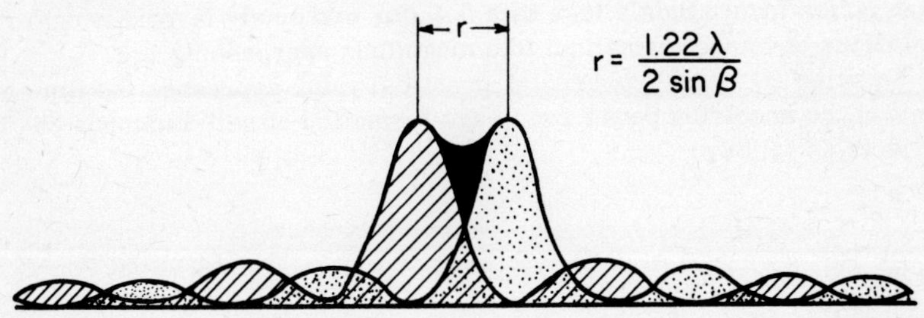


Figure 14.  The diffraction discs in the image plane overlap. The distance between their centers is equal to their radii. The increase of intensity in the area between the maxima, to which both diffraction discs contribute, is indicated by the black area in the diagram.

the object detail in the image. Under these conditions, a reasonable LIMIT OF THE RESOLVING POWER of the objective has been reached. The following equation is often quoted as expressing the relationship between the factors which limit the resolving power of the objective in the formation of images of self-luminous objects:

$$D_O = 1.22\lambda/2NA \quad \text{or:} \quad D_O = 0.61\lambda/NA \qquad \text{Equation (7)}$$

This equation states that the shorter the wavelength of the light forming the image and the higher the NA of the objective, the better the resolving power of the objective; that means the smaller is the magnitude of the finest object detail which can be reproduced in the image.

The resolving power of the objective is not as sharply limited as the equation indicates. It varies within small limits and depends on several factors. The capacity of an observer, for instance, to detect small differences of intensity is an individual factor. Furthermore, small residual aberrations due to the limited state of correction of the objective influence the variations of the intensity throughout the diffraction discs. Therefore, it has been suggested that the factor 0.61 should be replaced by 1.22c with the understanding that the magnitude of the new factor c varies from slightly less than 0.5 (for extremely favorable conditions of image formation) to a magnitude approaching 1.0. Equation (8) which has greater practical value expresses the limit of the resolving power for image formation of self-luminous objects as follows:

$$D_O = 1.22c\,\lambda/NA \qquad \text{Equation (8)}$$

An <u>increase</u> in resolving power of the objective can be achieved by using light of shorter wavelengths. For visual observation, limitations arise, however, because the eye is not sensitive to wavelengths shorter than about 400 nanometers. The ultra-violet range can be used for photomicrography if all of the optical components of the illumination system as well as the image-forming system are made of materials which transmit ultraviolet light. Furthermore, the immersion liquid, the slide to which the object is mounted and also the coverslip must transmit ultraviolet light.  The liquid commonly used for immersion objectives transmitting ultraviolet light is glycerin.

It is also possible to increase the resolving power by designing objectives for immersion media with refractive indices higher than that of the oil generally used.  At one time, objectives were available with mono-bromonaphthalene ($n_D$= 1.66) as immersion fluid.  These objectives had an NA of 1.60.  When they were used for image formation of transparent objects which had to be illuminated by transmitted light, special glass slides and coverslips with equally high refractive indices had to be used and the objects had to be embedded in media of equally high refractive index.  The practical advantages which could be derived from the use of this type of objective were not as great as had been expected and they have disappeared from the market.

Objectives for photomicrography in the ultraviolet region of the spectrum are still available and have been described in Volume 14, page 86.  In addition to their higher resolving power, they have advantages for image formation of nonself-luminous objects, those which are practically completely transparent within the visible light range but which show considerable variation of absorption of wavelengths within the ultraviolet range.  Photomicrographs revealing their structures in good contrast to the background can be taken by correct selection of the wavelength.

Figure 15 shows the relationship between resolving power, wavelength and numerical aperture.

F.  <u>VARIATION IN LIGHT INTENSITY THROUGHOUT THE AREA OF THE OBJECT AND THE IMAGE</u>:  As mentioned before, the intensity of the light emitted by a single, self-luminous object point is equal to the square of the amplitude of the emitted waves.  If there are two or more adjacent object points, emitting light of <u>different</u> intensities, these variations are reproduced in the image because the object collects from each object point an equal share of the total light intensity, and light waves from adjacent diffraction discs in the image plane cannot interfere with each other.

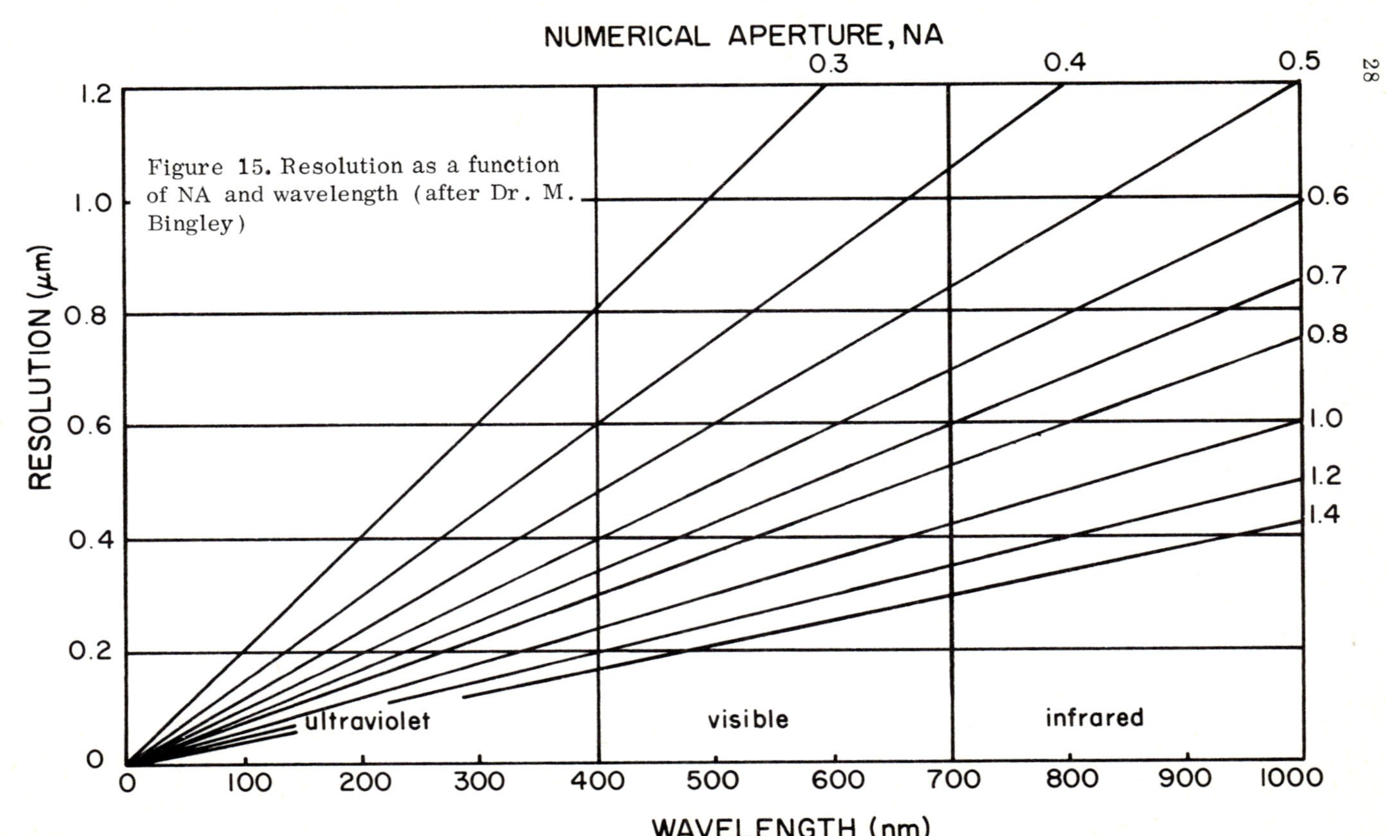


Figure 15. Resolution as a function of NA and wavelength (after Dr. M. Bingley)

A self-luminous object of finite size can be interpreted as consisting of an infinite number of self-luminous object points, and variations of the intensity throughout the area of the object are accurately reproduced in the image. The optical relation between the variations of the intensity throughout the object and image areas can be described as "optical fidelity of reproduction" which, in images of self-luminous objects, prevails in addition to "morphological resemblance" and is a characteristic of great practical significance.

G. DEPTH OF FIELD — AXIAL RESOLVING POWER: In the image formed by the objective, not only the contents of the object plane as defined geometrically are in optimum focus, but also object detail in planes within small, but a finite and measurable axial vicinity. An approach to the explanation of this fact can be based on certain geometrical optical aspects of image formation.

According to geometrical optics, the image of a "point" is also a point. Neither the limit of the resolving power nor the depth of field can be explained on the basis of geometrical optics because both factors are infinitely small.

Since, however, the image of a single object point, due to the physical nature of light, has finite lateral extension in all directions from the geometrical point of the image plane (the diffraction disc), geometrical relations between the object plane, the image plane and planes immediately adjacent to these planes in axial directions can be shown in a diagram (Figure 16).

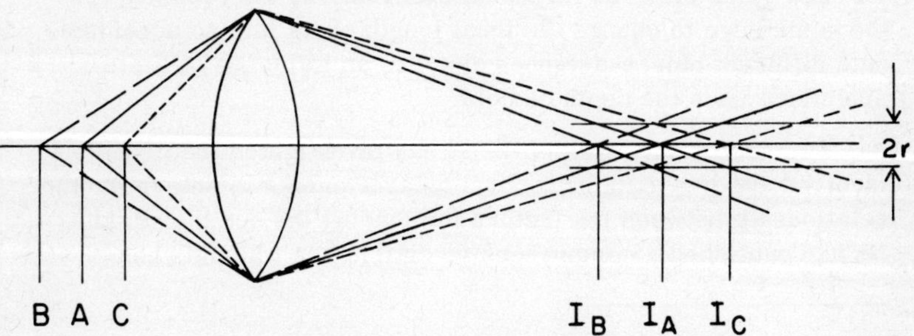


Figure 16. Depth of field

The image of object point A is at $I_a$ of the image plane and has the finite diameter 2r. The image of point B, slightly farther away from the objective than point A, is formed at slightly closer distance to the objective at $I_b$. From that image point, the "rays" diverge and in the plane of image point $I_a$ they occupy an area of the finite diameter 2r (which is the diameter of the diffraction disc of the image of point A).

The image of object point C, slightly closer to the objective than point A, is formed at slightly longer distance. In the plane of image point $I_a$ the rays from point C have not yet intersected in the geometrical image point, but occupy an area of the finite diameter 2r. Consequently, in the plane of the image of point A the images of points B and C are also in the best <u>physically obtainable</u> "sharpness" due to the physical nature of light.

This interpretation of the "depth of field" is not absolutely correct optically. The interference pattern which is the image of a single self-luminous object point has not only <u>lateral</u> extension in the image plane, but also <u>axial</u> extension to both sides of the image <u>plane</u>. The exact shape of this pattern in the axial direction is not the same as indicated by the geometrical optical interpretation.

The mathematical interpretation of the physical optical factors which limit the depth of field is very complicated. Not only the numerical aperture of the objective and the wavelength of the light forming the image influence the depth of field, but also the total magnification of the image, the refractive index of the object and other factors. In visual observation, the capacity of the human eye to change the focal length of its lens to accommodate different object distances also has an influence of finite magnitude upon the depth of field.

Professor M. Berek of Leitz has investigated the problems involved and has made numerous measurements to determine the relationship between the factors influencing the depth of field. He has published a comprehensive article[1] in which he has

---

[1] "Grundlagen der Tiefenwahrenhung im Mikroskop". Sitzungsberichte der Gesellschaft zur Beförderung der gesamten Naturwissenschaften zu Marburg <u>62</u>, 6 (1927).

derived an equation, based on theory as well as practice, from which the depth of field can be calculated when the NA of the objective, the total magnification and the refractive index of the object are known. Incidentally, he has also used the well chosen name AXIAL RESOLVING POWER for that range which is known under the name: depth of field.

Berek's equation as well as explanations and a diagram showing the range of the depth of field for various ranges of NA of the objective and magnifications are to be found in a booklet, published by Leitz.[1]

Berek's Equation (9) expresses the relationship between the factors which influence the axial resolving power as follows:

$$T_O = n_O[\, 4C\lambda/NA^2 + s\omega/NA(M)] \qquad \text{Equation (9)}$$

$T_O$ = Axial resolving power (depth of field)

$n_O$ = Refractive index of the object

NA = Numerical aperture of the objective

S  = Minimum distance of distinct vision (250 mm)

M  = Total magnification of the image

C  = An empirically determined factor of the magnitude, 1/8

$\omega$  = Another empirically determined factor of the magnitude 0.00136

$\lambda$  = Wavelength of light forming the image.

This equation is valid for visual observation and is based on image formation of an object under conditions of self-luminosity when the NA of the illumination is equal to that of the objective.

According to Berek's equation, the depth of field decreases with the square of the NA of the objective and the first power of the total magnification. When an objective of given NA is used

---

[1]Image-forming and Illuminating Systems of the Microscope, E. Leitz, Rockleigh, New Jersey 07647. The Microscope and its Application, (1970).

in combination with an ocular of 10X magnification, the depth of field is <u>greater</u> than when the same objective is used with an ocular of 20X magnification.

Approximate values for the depth of field have been calculated (Table II) with Berek's equation for certain ranges of magnifications and objectives of various NA, based on a selected value for the refractive index of the object of 1.45.

Table II

Depth of field ($\mu$m)

| Numerical aperture | MAGNIFICATION | | | | | | |
|---|---|---|---|---|---|---|---|
| | 25X | 63X | 100X | 200X | 400X | 1000X | 2000X |
| 0.08 | 308 | 160 | 124 | | | | |
| 0.16 | | 77 | 54 | 34 | | | |
| 0.22 | | | 32 | 21 | | | |
| 0.65 | | | | | 2.8 | 1.7 | |
| 1.25 | | | | | | 0.65 | 0.45 |

At low magnifications, the depth of field exceeds these values because the capacity of the observer's eyes to accommodate to various object distances influences its range to a greater extent than at high magnifications.

H.  <u>PRIMARY IMAGE FORMATION</u>: The formation of images of self-luminous objects has been called PRIMARY IMAGE FORMATION because the light waves emitted by the object proceed as expanding spherical wave surfaces through the object space and the limited portions which are collected by the objective undergo such changes that on emergence into the image space, they are again parts of spheres which proceed <u>directly</u> towards their new centers in the image plane. In the following chapter the formation of images of nonself-luminous objects is analyzed. Under certain illumination conditions, modifications of the wave surfaces in their passage from the object plane to the image plane are more complex.

IMAGE FORMATION OF NONSELF-LUMINOUS OBJECTS
UNIDIRECTIONAL ILLUMINATION

A.  UNIDIRECTIONAL AXIAL ILLUMINATION:  COHERENCE
IN OBJECT PLANE: An optical system for illumination of
transparent objects by transmitted light was described in
Volume 14, Chapter 4 and the lightpath through this system, the
"Köhler illumination system" was shown there (Figure 28 on
page 48).

The image of the light source is formed in the lower focal
plane of the condenser. This is also the plane of the aperture
(iris) diaphram. When it is closed as far as possible (theoretically to a diameter  equal to a diffraction disc) only light originating in a single point of the light source can pass through this
iris diaphram. In their passage through the condenser, the expanding wave surfaces undergo modifications and emerge to pass
through the object plane as plane wave surfaces, illuminating the
object with light of one single direction, that of the optical axis
(unidirectional axial illumination). Coherence prevails throughout the area of the object.

When there is no object in the object plane, the plane wave
surfaces enter the objective and, in their passage through it,
their shapes are changed again so that they emerge as converg-
ing spherical wave surfaces which proceed to form an image of
the single point of the light source and also of the aperture diaphram of the condenser in the back focal plane of the objective.

It is, of course, impossible to close the iris diaphram of
the condenser so much that its diameter is equal to that of a
single diffraction disc. In practice, the image formed in the
back focal plane of the objective is that of a small circular area
of the light source, limited by the image of the rim of the aperture stop of the condenser. It can be observed when an intermediate optical system (Bertrand lens or centering telescope for
phase contrast illumination) is interposed in the lightpath or
when the ocular is removed for direct viewing. The lightpath
from the lower focal plane of the condenser through the object
plane to the back focal plane of the objective is shown in Figure
17a.

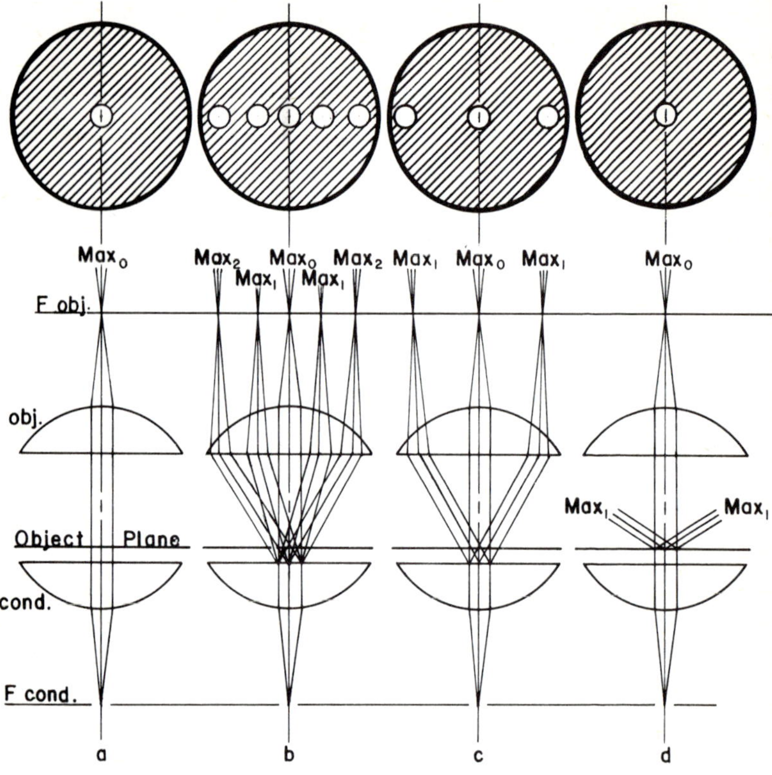


Figure 17a: Unidirectional axial illumination.  No object in the
object plane.

b: Same illumination conditions.  The structure of a
grating produces interference maxima.  $Max_1$ and
$Max_2$ in addition to $Max_0$, showing angles of inclination symmetrical to $Max_0$, are collected and
transmitted by the objective.

c: Because of the finer structure of the grating only
$Max_1$, in addition to $Max_0$, is collected.

d: The grating structure is so fine that only $Max_0$ is
collected by the objective and the structure is not
resolved.

When the object is a single, <u>non</u>self-luminous point, <u>e.g.</u>, a hole of extremely small diameter in a metallic film on a glass slide — and this hole is illuminated with unidirectional, axial illumination, it becomes the center of diffracted waves which proceed as expanding spherical wave surfaces towards the objective. There is no difference between the formation of the image of a self-luminous or a nonself-luminous <u>single</u> point.

Image information of <u>nonself-luminous objects of finite size</u> under conditions of <u>unidirectional illumination and coherence in the object plane</u> was analyzed mathematically and interpreted by Professor Ernst Abbe of Zeiss. It is known as the ABBE THEORY.

B. <u>THE ABBE THEORY: RESOLVING POWER AND IMAGE CHARACTER</u>: For the analysis of the optical phenomena produced by the structure of the object under condition of unidirectional illumination, Professor Abbe selected objects of known periodic structure, <u>e.g.</u>, two parallel slits, gratings and other objects.

When a grating is placed in the object plane, in the lightpath of the coherent plane wave surfaces, its structures cause the emission of diffracted waves which interfere with each other and produce interference maxima and minima. The path of the points of intersection of two sets of diffracted waves of <u>equal phase</u>, originating in two adjacent "slits" of the grating and interfering <u>without</u> pathlength difference, is a straight line in the same direction as the illumination, parallel to the optical axis. Since the grating consists of a great number of adjacent slits or lines, there are many points of intersections of maxima without pathlength difference from adjacent points and all of them proceed parallel to each other in the direction of the optical axis. As they emerge from the objective, they converge and form an image of a single point of the light source in its back focal plane.

Also the points of intersection of diffracted waves, interfering <u>with</u> a pathlength difference of one <u>full</u> wavelength ($Max_1$), proceed parallel to each other, but at an angle of inclination to the optical axis.

Depending upon the NA of the objective and the magnitude of the grating constant (the space between adjacent lines of the grating), interference maxima of higher orders may also be collected by the objective. Each forms an additional image of the same point of the light source in the back focal plane of the objective (Figure 17b). The maximum $\text{Max}_0$ is in the back focal plane and additional maxima of higher orders are to the right and left, at equal distances, respectively, from it.

In Figure 17b, the grating constant is of such magnitude that maxima of the first and second order are collected by the objective. Since these additional images originate in one single point of the light source, light proceeding from them toward the image plane is coherent and once again these waves interfere with each other. In the image plane, an interference pattern is produced in which the light intensity changes rather abruptly from a maximum to a minimum, depending upon the pathlength difference from any single image point to the coherent light centers in the back focal plane of the objective. In this pattern, the periodic structure of the grating is reproduced with the distance between adjacent lines magnified according to the laws of geometrical optics. The image has "morphological resemblance" to the object, but there is no "optical fidelity of reproduction" because the light at any single point of the image does not originate in a single object point, but is the result of interference of light waves from at least two adjacent centers in the back focal plane of the objective.

As mentioned before, it is impossible to close the aperture diaphram of the condenser so much that light from only one diffraction disc can pass through it. The images, formed in the back focal plane of the objective, are those of a circular area of very small diameter of the light source. However, to each selected point within one of these images, a point can be coordinated within the adjacent image (interference maximum) which is traceable to the same point of origin in the light source, and light from such respective points is coherent and capable of interference. The total effect of all interferences is the pattern produced in the image plane.

According to Equation 1 (page 8), the smaller the grating constant, the greater is the angle of inclination to the optical axis at which the first interference maximum (and the adjacent maxima of higher orders) proceeds. When the grating constant

D is of such magnitude that the first interference maximum proceeds at an angle of inclination, equal to that of highest inclination at which light can proceed to be collected by an objective of given NA, the two maxima of the first order appear at the periphery of the aperture stop in the back focal plane of the objective, separated from $\text{Max}_0$ by a distance equal to the <u>radius</u> of that aperture stop (Figure 17c).

When the grating constant is still smaller, the first interference maxima proceed at such high angles of inclination that they are not collected by the objective (Figure 17d). There is only one <u>single</u> center of light in the back focal plane of the objective — that of $\text{Max}_0$. The optical conditions are the same as those shown in Figure 17a when there is no object in the object plane. The structures of such gratings are not "resolved".

The limit of the resolving power of the objective for unidirectional, axial illumination has been reached when the following relationship prevails between the distance D, separating adjacent structures of the object (the grating constant), the NA of the objective and the wavelength $\lambda$ of the light, forming the image:

$$D = \lambda/NA \qquad \text{Equation (10)}$$

When the closed aperture stop of the condenser is <u>laterally</u> displaced in its lower focal plane, its images in the back focal plane of the objective also undergo lateral displacements. One of the first-order maxima proceeds at an increased angle of inclination and is not collected by the objective. The other maximum of the same order, however, proceeds at a lower angle of inclination and may now be collected by the objective, in addition to $\text{Max}_0$. There are now <u>two</u> adjacent maxima collected by the objective and light from them produces an interference pattern in the image plane in which the structures of the grating are resolved (Figure 18a).

In order to make lateral displacements of the aperture diaphram of the condenser possible, Professor Abbe designed the ABBE ILLUMINATION APPARATUS. It is provided with facilities for lateral displacement of the (closed) iris diaphram and for rotating it around the optical axis to change the azimuth of the oblique illumination to the most favorable angle with respect to the direction of linear periodic object structures.

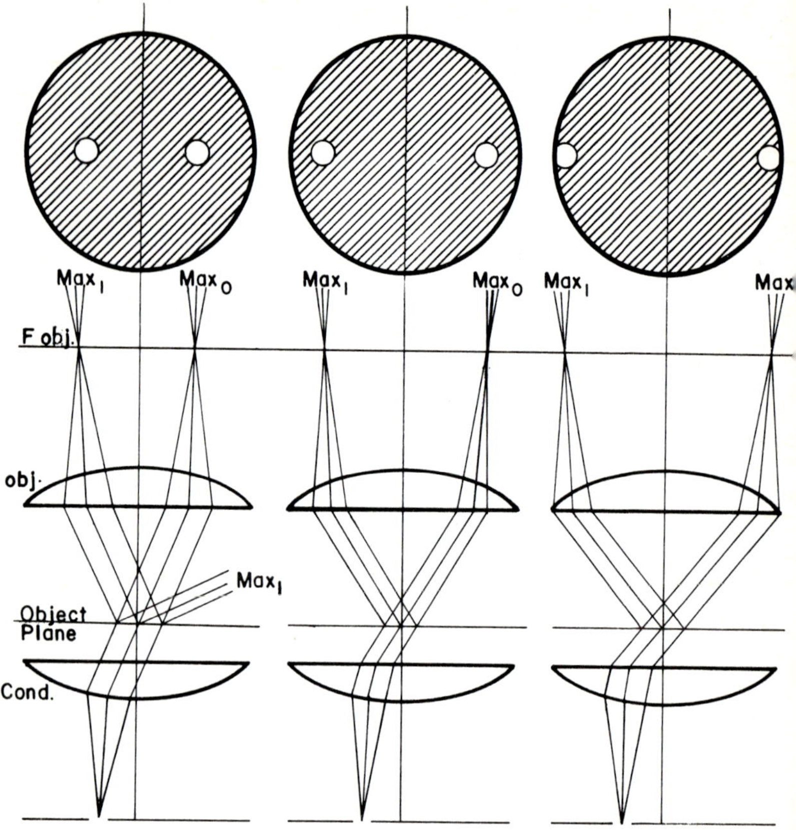


Figure 18a:  Slightly oblique illumination.  $Max_0$ and $Max_1$ are collected by the objective, separated by a distance equal to the radius of the aperture stop.

b:  The distance between adjacent maxima is 3/4d.

c:  The limit of resolving power for oblique illumination.  The distance between adjacent maxima is equal to the diameter of the aperture stop.

Under conditions of oblique unidirectional illumination, it is possible to resolve finer periodic structures. In Figure 18b, the grating structure is so much finer that adjacent maxima $Max_0$ and $Max_1$ in the back focal plane of the objective are separated by a distance equal to 3/4 of the <u>diameter</u> of the objective aperture stop.

When the oblique illumination proceeds at an angle i to the optical axis and the refractive index of the medium between condenser and object plane is n, the relationship between the factors limiting the resolving power is:

$$D = \lambda/n \sin i + NA \qquad \text{Equation (11)}$$

In some publications on the subject, the factor n sin i is quoted as $NA_i$, the numerical aperture of the illumination. This is misleading because it can be interpreted to the effect that the equation is valid for illumination of the object by a <u>full cone</u> of light, proceeding continuously in directions from that of the optical axis to a maximum inclination, corresponding to $NA_i$. Although mathematically the two factors are equal, Equation (11) has been derived for <u>unidirectional</u> oblique illumination. Its validity for multidirectional illumination requires additional investigations, not only in regard to the resolving power under those conditions, but also with regard to the optical character of the image.

When the grating structure is so fine that, under unidirectional oblique illumination at maximum angle of inclination to the optical axis (which the objective of given NA can collect) the interference maxima $Max_0$ and $Max_1$ in the back focal plane of the objective are separated by a distance equal to the <u>diameter</u> of its aperture stop, the resolving power is twice as high as that prevailing for unidirectional <u>axial</u> illumination and is expressed by the equation:

$$D = \lambda/2 \, NA \qquad \text{Equation (12)}$$

The optical conditions for unidirectional oblique illumination of maximum inclination and for grating structures about equal to the limit of the resolving power of the objective, are shown in Figure 18c.

Whether or not the optical character of the image formed under conditions of unidirectional illumination is advantageous depends upon the optical character of the <u>object</u>.

In the introduction to Volume 14 attention was drawn to one essential prerequisite for image formation: the object must differ optically sufficiently from its surroundings. When the object possesses these optical differences inherently or when they have been created by special preparation for microscopical observation, <u>e.g.</u>, by staining, an image of the optical character described as "optical fidelity of reproduction" is most advantageous. If, however, the increase of optical differences by object preparation is not feasible, image formation under conditions of unidirectional illumination with sharp transitions from brightness to darkness (increased contrast) may be required to reveal periodic object structures in the image.

The photomicrograph in Figure 19, showing the periodic

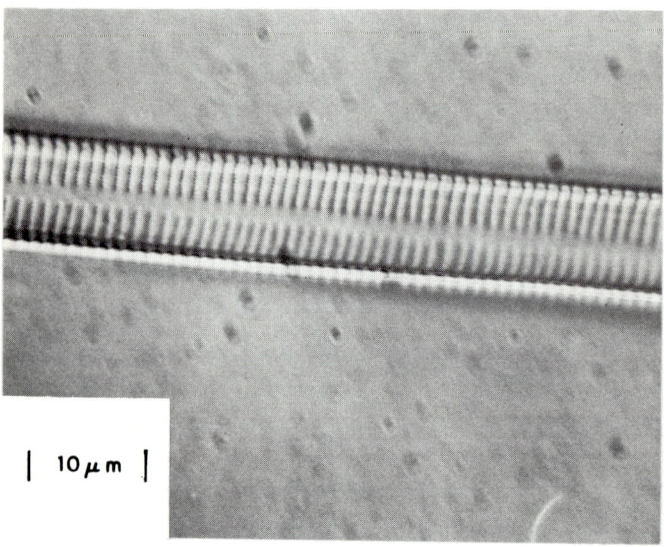


Figure 19. Amphipleura pellucida photographed with extremely oblique illumination.

structures of the diatom <u>AMPHIPLEURA</u> <u>PELLUCIDA</u>, was taken under conditions of extremely oblique illumination.* If the same object had been photographed under conditions of multidirectional illumination with an NA equal to that of the objective, these structures would still be resolved, but the contrast in the image would have been reduced to such an extent that they could hardly be seen.

Increase of contrast in the image can be obtained also with unidirectional <u>axial</u> illumination of the object. Unfortunately, under these conditions, the resolving power is substantially reduced. Under conditions of oblique illumination with increased contrast, the increased resolving power depends on the azimuth of the illumination in respect to periodical structures of the object.

Fortunately, increased contrast in the image can be achieved without loss of resolving power, even for objects with irregular structures differing from their surroundings mainly with respect to their refractive index and even when these differences are quite small. This is achieved, for instance, under conditions of PHASE CONTRAST ILLUMINATION, the subject for another volume of this series.

C. <u>SECONDARY IMAGE FORMATION</u>: Image formation under conditions of <u>unidirectional</u> illumination on which the Abbe Theory is based, has been called SECONDARY IMAGE FORMATION in contrast to primary image formation as described for self-luminous objects.

In the passage of light from the object plane under conditions of <u>unidirectional</u> illumination, there are actually <u>two</u> stages of image formation. The first is the formation of additional (coherent) images of very small areas of the light source in the back focal plane of the objective. The formation of these images is due to <u>interference</u> of <u>coherent</u> light waves, diffracted by the <u>object structures</u>. The second image formation is that of the object structures in the image plane, which is due to <u>interference</u> of <u>coherent</u> light waves proceeding from the <u>images of the</u>

---

*By changing the azimuth of the oblique illumination and using a sufficiently short wavelength, the parallel "lines" of the structure can be resolved into rows of "dots".

light source in the back focal plane of the objective.

D. EXPERIMENTS WITH THE ABBE DIFFRACTION APPARATUS: A series of very instructive demonstrations can be made with the Abbe Diffraction Apparatus to show the relation between the coherent light centers in the back focal plane of the objective and the "contents" of the interference pattern produced in the image plane.

The apparatus consists of several accessories which can be attached to a transmitted light microscope. An objective of low magnification and suitable numerical aperture is attached to an intermediate collar which is provided with a slot. Several interchangeable sliders (for a variety of demonstrations) can be inserted into the slot. One is equipped with an iris diaphram for reduction of the numerical aperture of the objective. Another slider has a circular hole with a recess into which one of several metal discs can be placed. These discs have open spaces of several shapes so that selected interference maxima can be blocked from passage to the image plane. The objective with the slotted collar is attached to the microscope.

Several test objects with periodic structures are mounted on one common object slide. One is a combination of two linear rulings. The distances between adjacent rulings of the coarser structure are twice as wide as those of the finer ruling. A series of photomicrographs of this and other test objects were taken with a microscope with a built-in illumination system for Kohler illumination. Because of the low magnification and the large field of view, the front lens of the condenser was swung out of the light path so that the iris diaphram, which acted as a field stop at high magnification, performed as an aperture stop for the illumination. The microscope was also equipped with a device for observation of the back focal plane of the objective (Zeiss OPTOVAR).

The photomicrograph of Figure 20 shows the two rulings, so aligned that each occupies one half the field of view. Before taking the next photomicrograph (Figure 21) the object was moved so that only the coarse rulings were visible in the field of view. After taking this photomicrograph, the film was not advanced. The Optovar was interposed in the lightpath (a Bertrand lens can also be used) and the image of the (closed) aperture stop

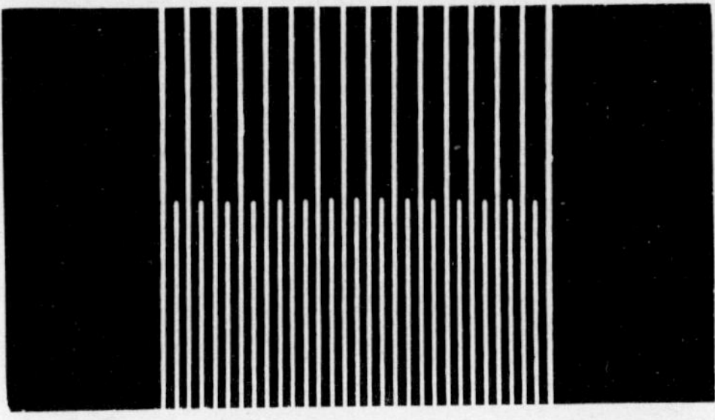

Figure 20. Coarse and fine rulings photographed from the Abbe diffraction test plate.

of the illumination system was focused. A second exposure was
taken on the same film frame. For this and the following photomicrographs, a green interference filter was interposed in the
lightpath. The maxima of higher orders, formed with green
light, appear as sharp as the maximum without pathlength difference. Without the filter in the lightpath, the maxima of higher orders, formed by light of all of the wavelengths of the visible spectrum at increasing distances from the maximum of zero
order, would have produced colored elongated "blurs". The
double exposure of Figure 21 shows the ruling and the interference maxima of zero to fourth orders.

The next photomicrograph was taken under similar conditions with only the fine ruling in the lightpath (Figure 22). The
interference maximum $Max_1$ of the fine ruling has the same distance from $Max_0$ as the maximum $Max_2$ of the coarse ruling. A
third photomicrograph was taken under similar conditions,
(Figure 23) with both rulings visible within the field of view, as
in Figure 20. Maxima $Max_1$ and $Max_3$ of the coarse ruling appear with lower light intensity than $Max_2$ and $Max_4$ because
these latter maxima coincide with $Max_1$ and $Max_3$ of the fine
ruling.

The slider with the iris diaphram was now inserted into the
slot of the intermediate collar and was closed until only the two
(symmetrical) maxima $Max_1$ of the coarse ruling could pass
through it, in addition to $Max_0$. All of the maxima of the fine
fulings, except $Max_0$ were blocked by the partly closed iris

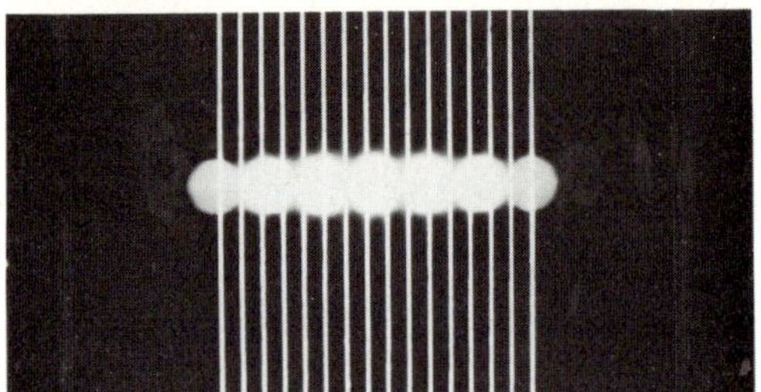

Figure 21.   Superimposed diffraction pattern by, and photomicrograph of, coarse rulings.

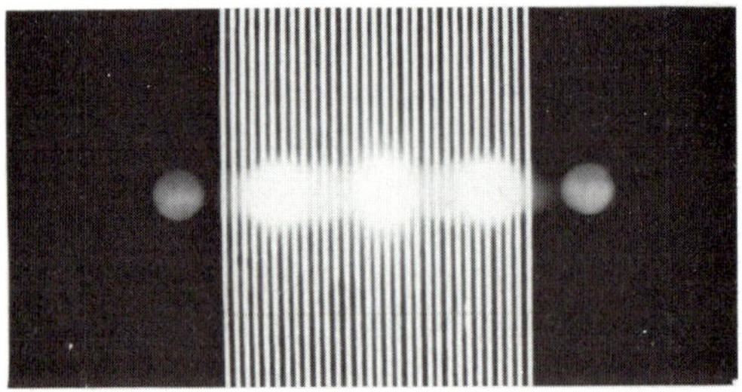

Figure 22.   Superimposed diffraction pattern by, and photomicrograph of, fine rulings.

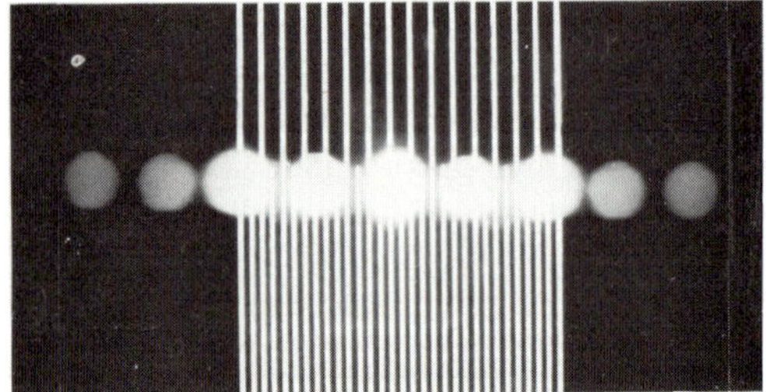

Figure 23.   Superimposed diffraction pattern by, and photomicrographs of, both fine and coarse rulings; $Max_2$ and $Max_4$ of the coarse ruling coincide with $Max_1$ and $Max_2$ of the fine ruling.

diaphram.  Under these conditions, only the structure of the coarse ruling is resolved.  (See Figure 24).

The condenser was next slightly decentered for oblique illumination.  The iris diaphram of the slider, acting as an aperture stop of the objective, was partly closed and the object was moved so that only the coarse ruling was visible.  On the double exposure of the photomicrograph (Figure 25), only <u>two</u> adjacent interference maxima are seen.  The structure is still resolved, but the relative widths of the white lines and the black spaces between them are quite different from those of Figure 20, when as many as four maxima to each side of $Max_0$ were participating in the formation of the image.  The lower the number of adjacent interference maxima participating in the formation of the image, the lower is the morphological resemblance between object detail of a magnitude near the limit of the resolving power and its reproduction in the image.  The <u>geometrical</u> relationship between the <u>distances</u> separating adjacent object structures  is still correctly reproduced in the image.

Another slider was interposed in the lightpath, after a disc with several parallel slits was inserted into the recess of its circular hole.  The widths of these slits and their distances from each other were of such magnitudes that the first and third maxima of the coarse ruling were blocked and did not participate in the formation of the image.  The object was moved to the same position as in Figure 20.  Since the optical phenomena, produced exclusively by the coarse ruling ($Max_1$ and $Max_2$), were prevented from passing to the image plane, the effect in that image is the same as that produced by the fine ruling.  The coarse ruling appears to be as fine as the fine ruling.  In other words, lines appear in the image which are not in the object.  This is shown in Figure 26.

Similar "dramatic" effects can be produced by another test object.  It consists of bright "points" with opaque spaces between them arranged in lines within a square area (Figure 27).  This photomicrograph also shows the interference maxima in the back focal plane of the objective.  A metal disc with only one slit was placed in the recess of the slider and was so oriented that only three maxima in a direction, parallel to the side line of the square, could pass through the slit.  In the image, the object

Figure 24. An objective of lower NA was used for this and the next two photos to obtain greater lateral separation of the Maxima. Only $Max_0$ and $Max_1$ are collected by the objective. The structure of the fine ruling is not resolved.

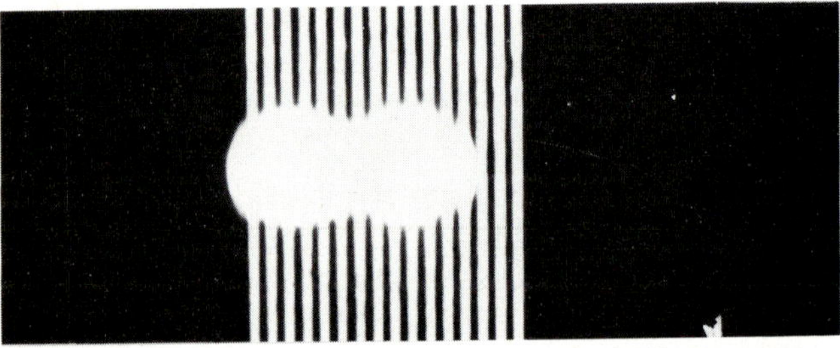

Figure 25. Slightly oblique unidirectional illumination. Only $Max_0$ and <u>one</u> $Max_1$ are collected by the objective. The structure of the coarse ruling is still resolved, but the white lines are too wide.

Figure 26. Unidirectional axial illumination. $Max_1$ and $Max_2$ of the coarse ruling are blocked by a multislip aperture stop. The coarse ruling appears as fine as the fine ruling.

appears as parallel lines, parallel to the edge of the square. By rotating the slider to orient the slit in a diagonal direction, the object appears to consist of lines in diagonal directions. (Figures 28 and 29). The next set of double exposures of the selected test object was taken after the aperture iris diaphram in the back focal plane of the objective had been closed so much that only $Max_0$ could pass through it. It is visible in the center of Figure 30. There is no other interference maximum with which light from $Max_0$ can interfere and the structure of the object is not resolved.

Before removing the test object, the green filter was removed from the lightpath so that all of the colors of the spectrum could produce interference maxima at increasing distances from $Max_0$. They appear as elongated blurs (Figure 31).

A disc with a ring-shaped transparent area was placed near the plane of the (fully open) aperture diaphram of the condenser. The entire area of the "annulus" was illuminated. Its multiple images in the back focal plane of the objective diffracted by the square area of small dots overlap and produce an interesting pattern (Figure 32).

These demonstrations with the Abbe Diffraction Apparatus prove that in the formation of the intermediate image of an object with periodic structures, under conditions of unidirectional illumination and coherence in the object plane, the variations of intensity in the image plane are caused by interference of light waves proceeding <u>from the coherent centers in the back focal plane of the objective</u>. These centers are images of the light source. By selecting a restricted number of these coherent centers for participation in the formation of the intermediate image and preventing light from other centers from reaching the image plane, object detail appears in the image plane which does not exist in the object plane.

The Abbe theory in its original form is restricted to image formation of objects with periodic structures, illuminated by unidirectional light. Although these illumination conditions are rarely used by practicing microscopists, the great value of this theory for further developments in microscopy is to be found in the systematic interpretations of the physical-optical conditions

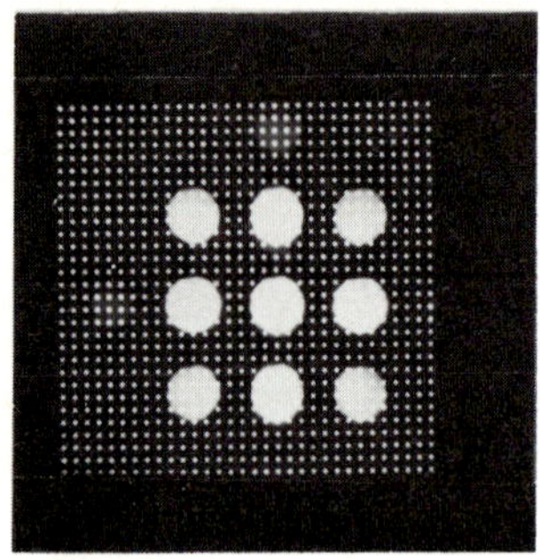

Figure 27. Square area with small "dots" and superimposed pattern of interference maxima in back focal plane of the objective.

Figure 28. Only maxima in one horizontal line are transmitted by single slit aperture stop. The square area of small dots now seems to consist of parallel vertical lines.

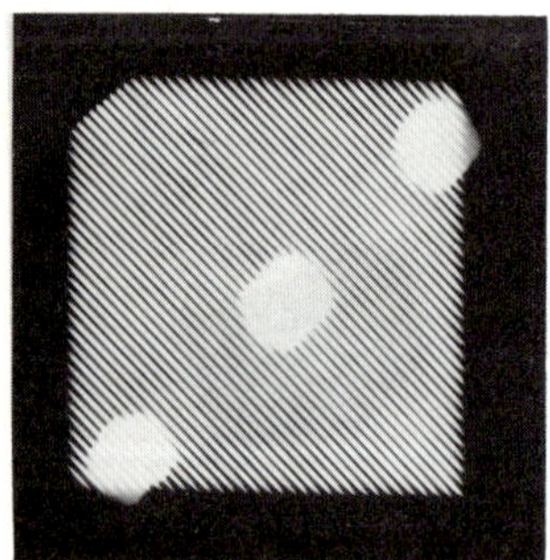

Figure 29. The slit-shaped aperture stop has been rotated 45° and the square area of small dots now appear to be diagonal lines.

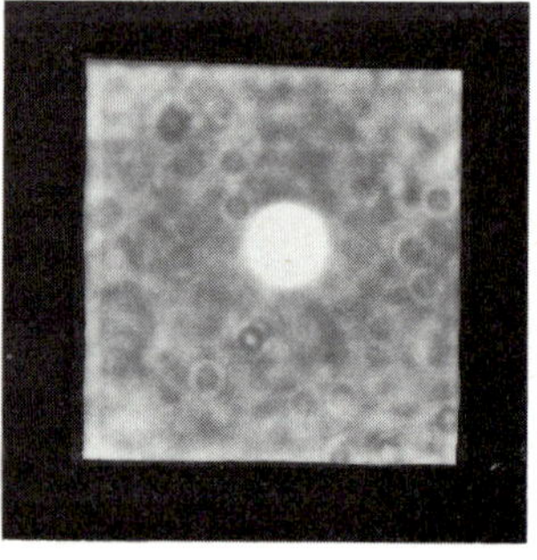

Figure 30. Only $Max_0$ is transmitted by the almost completely closed aperture iris diaphram hence the structure of the object is not resolved.

Figure 31. The monochromatic green filter has been removed. The spectral colors of white light form maxima at distances increasing with wavelength and appear as elongated "blurs".

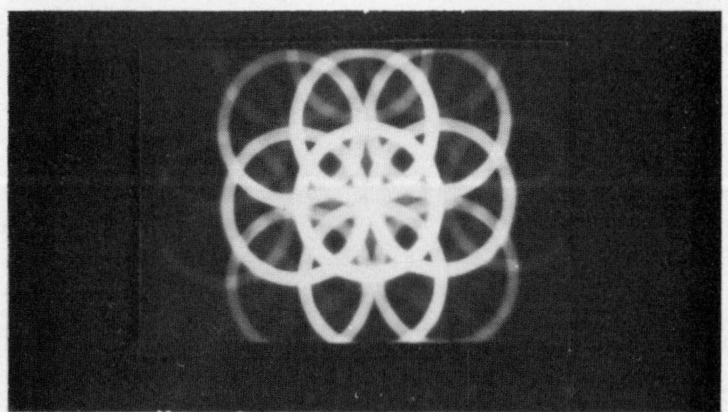

Figure 32. A ring-shaped aperture stop, like a phase annulus, was used in the lower focal plane of the condenser, producing this "dramatic" pattern of diffraction images in the back focal plane of the objective. The square area of small dots (Figures 25-29) was used to produce the diffraction images.

of image formation and the differentiation between primary image formation of self-luminous objects and secondary image formation of nonself-luminous objects. Professor Abbe's work paved a way for further systematic analyses and lifted microscopy from a state of practical experimentation to an exact science.

# IMAGE FORMATION OF NONSELF-LUMINOUS OBJECTS
## MULTIDIRECTIONAL ILLUMINATION

**A. EQUIVALENCE TO SELF-LUMINOSITY, INCOHERENCE IN THE OBJECT PLANE:** The optical characteristics of images of self-luminous objects are advantageous for many nonself-luminous objects. Stained histological sections, for instance, embedded in media of practically the same refractive index, differ optically from their surroundings mainly with respect to <u>variations</u> of light <u>absorption</u>. Not only does the percentage of light absorption vary from one point to another for one given color, but for another color the percentage may vary to a different degree. In the images of this type of object, these variations should be reproduced with optical fidelity of reproduction.

The geometrical optical aspects of the performance of one illumination system — <u>and not the only one</u> — which creates optical conditions of equivalence to self-luminosity in the object plane were briefly described on page 57 of Volume 14, under the name of Nelsonian (critical) illumination. In describing the physical optical aspects of the performance of this system, the descriptive adjective "critical" will be used for reasons which will be unfolded in the course of the following explanations.

**B. PHYSICAL OPTICAL ASPECTS OF NELSONIAN (CRITICAL) ILLUMINATION:** Two optical requirements are generally stated in the descriptions of this illumination method:

1. A real image of the light source must be formed in the plane of the object.

2. The "back lens" of the objective must be filled with light.

To fulfill the first requirement, a completely homogeneous light source must be used so that the amplitudes of the light waves emitted by each one of its "points" are of equal magnitude and the intensity of the light does not vary throughout its image in the object plane. Under these conditions, incoherence prevails throughout the field of view, down to an area equal to the diameter of a diffraction disc — the image of a single point of a light source.

To fulfill the second requirement, a condenser must be used which has as high an NA as the objective and is equally well corrected for aberrations.  Under these illumination conditions, the radii of the diffraction discs are equal to the limit of resolving power of the objective.  Each "point" in the object is illuminated by light originating in a single point of the light source.  From each object point, spherical wave surfaces proceed toward the objective, and light waves from adjacent object points, separated from each other by a distance equal to the limit of the resolving power of the objective, cannot interfere with each other because of incoherence which prevails down to that limit.

The amplitudes of the light waves proceeding from the object plane to the objective vary because of variations of absorption of light in the object and these variations are faithfully reproduced in the image.  The optical character of the image is that of optical fidelity of reproduction and the object has assumed the optical properties of equivalence to self-luminosity.  The limit of resolving power is the same as that expressed in Equation (8; page 26).

Evidently, these optical characteristics of the image were found to be advantageous by microscopists who observed, predominately, objects with irregular structures which were stained or had inherent colors.  That is, undoubtedly, the reason why Nelson advocated "critical illumination" and why many <u>practicing</u> microscopists used this illumination method.

Homogeneous light sources fulfilling the requirements for critical illumination were difficult to find.  It became a practice to focus the image of the "edge" of the light source as described in Volume 14, page 57, with subsequent readjustment of the mirror of the substage so that the image of the central area of the light source filled the field of view.  When the intensity of the light throughout the field of view showed too much variation, the condenser was moved slightly in the direction of the optical axis (throwing the image of the light source out of focus) to make the intensity throughout the field more uniform.  When this is done, the "critically" focused images of the points of the light source — the diffraction discs — are not in focus in the object plane and the adjective "critical" is not justified for a description of this method of illumination.

It is, however, a fact of great practical significance that under these changed illumination conditions keen observation of the image revealed that neither the optical character of the image nor the resolving power is changed to any detectable degree. Optical fidelity of reproduction still prevails throughout the area of the field of view.

C. <u>"AFOCAL" ILLUMINATION</u>: In the further course of time, heterogeneous light sources were used with a ground glass interposed in the lightpath to make the light intensity throughout the field of view uniform. Many early microscopists, who used their microscopes during the daytime, made use of the light from the "blue sky" to illuminate the object. In order to increase the light intensity in the object plane, the concave mirror of the substage was used. When they continued their observations in the evening, they had to use artificial light sources and, due to their lower color temperatures, the colors of the object appeared changed. In order to correct this, a blue filter was interposed in the lightpath to make the light similar to that of daylight. Filters of this type are still used extensively. When the illumination system does not form any image of the light source, it should be called AFOCAL.

Evidently, in practice, no changes of the optical character of the image or the resolving power were detected, as long as the back lens of the objective was still filled with light, <u>i.e.</u>, the NA of the illumination was equal to that of the objective. There is still incoherence in the object plane down to the limit of the resolving power of the objective and light waves from adjacent objects cannot interfere with each other.

D. <u>MULTIDIRECTIONAL KÖHLER ILLUMINATION</u>: When Köhler illumination is used and the aperture iris diaphram of the condenser is opened as far as possible, the object plane is traversed by an infinite number of coherent plane wave surfaces, proceeding in directions from that perpendicular to the optical axis to that perpendicular to the highest angle of inclination to it, at which light can proceed to be collected by the objective.

These plane wave surfaces also proceed at all possible azimuths. The NA of the illumination is equal to, or greater than, that of the objective. Also under these illumination conditions, incoherence prevails throughout the field of view in the object plane and the optical character of the image is that of optical fidelity of reproduction with unchanged resolving power and equivalence to self-luminosity.

To summarize, practical experience has shown that the optical character of the image, formed under conditions of critical, afocal or Köhler illumination, is unchanged as long as the numerical aperture of the illumination is as large as that of the objective. The formation of "critically" focused images of the points of the light source in the object plane is not "per se" an essential requirement for avoiding interferences of light waves proceeding from adjacent object points toward  the objective. The location of the image of the light source with respect to the object plane does not influence the resolving power or the optical character of the image.

In view of this, the Köhler illumination system is to be preferred because heterogeneous light sources can be used and, with the aid of a correctly performing field iris diaphram, the alignment of the illumination system is made easy. Furthermore, the detrimental effects of glare can be reduced.

E. <u>MULTIDIRECTIONAL ILLUMINATION WITH REDUCED NUMERICAL APERTURE</u>: Practicing microscopists generally do not use the microscope under conditions where the NA of the illumination is equal to that of the objective. They have found that the quality of the image can be improved by slightly closing the aperture stop of the condenser.

In some books on microscopy, specific recommendations are to be found for the reduction of condenser NA. Definite ratios between the NA of the illumination and that of the objective are recommended, for instance, 9/10, 4/5, 3/4 etc. These recommendations are generally made without detailed explanations of the cause of the improvement of the image quality. The fact that <u>fixed</u> ratios are recommended indicates that the reasons for improvement may not have been known to the authors. Actually there is <u>no</u> <u>fixed</u> ratio of the two numerical apertures at which optimum image quality prevails, since the optical factors <u>vary</u> not only from one object to another, but also from one objective to another.

One way to improve the image quality under conditions of

reduced NA of the illumination is the <u>reduction of glare</u> as described on page 56 of Volume 14.  The intensity of glare, due to partial reflections, increases with the angle of inclination to the optical axis at which light passes repeatedly from one medium into another.  By reducing the NA of the illumination, a corresponding reduction of the detrimental effect of glare is achieved.

Another possible reason for improved image quality may be the limited degree of correction of aberrations of the selected condenser system.  In Chapter 5 of Volume 14, the correction of spherical aberrations was described and it was mentioned that a state of correction can be achieved by relatively simple means but which leaves small residual amounts of spherical aberration uncorrected, particularly throughout the peripheral zones of the lens system.  When this light is prevented from reaching the object plane by partial closing of the aperture stop of the condenser, the general performance of the illumination system is improved.

There is still another reason for the improvement of the image quality under conditions of multidirectional illumination with reduced NA.  The best approach to an explanation of this is by referring again to the optical conditions of truly "critical" illumination.  A reduction of the NA of the illumination causes an increase in the diameters of the diffraction discs which are the images of single points of the light source.  Throughout a relatively large circular area of a diffraction disc coherence prevails.  When the diameter of that area is greater than the limit of the resolving power of the objective, light waves from adjacent object points, separated by a distance only slightly larger than the limit of its resolving power, can still interfere with each other, resulting in greater contrast in the image.  This increased contrast, however, is limited to object detail very close to the limit of the resolving power, whereas the general character of the image is still that of optical fidelity of reproduction. The diffracted light waves proceeding from these object points can still pass through the entire NA range of the objective, including the peripheral zones through which the light illuminating the object cannot pass.  Therefore, the resolving power of the objective is still higher than that which would prevail if the NA of the objective had been reduced to match the reduced NA of the illumination.

It is well to consider the validity of Equation 11, page 39 (which was originally derived for conditions of <u>unidirectional oblique illumination</u>) for conditions of <u>multidirectional illumination of reduced NA</u>. As far as resolving power is concerned, it is <u>the same</u> for unidirectional oblique illumination proceeding only at the single direction of angle i of inclination to the optical axis and multidirectional illumination with a full cone of directions from plus i to minus i, but the <u>optical characters</u> of the two images <u>differ</u> from each other. When unidirectional oblique illumination is used, the intensity throughout the image varies fairly abruptly from brightness to darkness. When multidirectional illumination is used, the general optical character of the image is that of optical fidelity of reproduction, almost down to the limit of the resolving power.

Also with Köhler illumination, the contrast in the image between adjacent object detail can be improved by slight reduction of the NA of the illumination, indicating that also under these illumination conditions there is a slight increase of the diameter of the circular areas of the object plane throughout which coherence prevails.

The exact magnitude of the reduction of the NA of the Illumination for production of images of optimum quality depends on <u>several</u> optical factors: character of the object and corrections of the objective. By the first we mean generally the degree of inherent contrast and by the latter we mean achromat <u>vs</u> planachromat <u>vs</u> apochromat <u>vs</u> planapochromat.

With the usual microscopic object and a moderate difference in refractive indices of object and mounting medium some reduction in the NA of the illumination will be of benefit. For a given $\Delta n$ there will, in fact, be an optimum setting of the NA such that sufficient contrast results without undue loss of resolving power. A difference in indices of at least 0.10 units, and better still, 0.15–0.20 units will usually allow a maximum NA, <u>i.e.</u>, equal to the objective NA.

The higher the degree of correction of the objective and, to a somewhat lesser extent, of the substage condenser the higher the NA of the illumination that can be used with a given degree of object contrast ($\Delta n$). Planapochromats and highly corrected condensers should therefore be used for the study of objects in a

medium of closely similar refractive index.  One can also use
phase contrast in these cases to improve contrast and thus main-
tain a reasonable NA with correspondingly higher resolving power.

It should be, in any case, obvious that the NA of the illumination should be set by observing the image rather than the objective back focal plane.  If, however, the best image quality is
obtained when the substage aperture is closed more than, say,
halfway steps should be taken to improve the situation.  First
perhaps, by remounting the specimen in a medium of substantially different refractive index.  If this is not practical, then interference or phase contrast should be used or, if your budget per-
mits, recourse to planapochromats will greatly improve the
image quality.

The critical reduction of the NA of the illumination, based
on keen observation of the image and evaluation of its quality, is
the most important single step in adjusting the optical system of
the microscope for optimum image quality.  If the aperture stop
is closed too much, the reduction of the resolving power and the
change of the optical character of the image become unacceptable.
If the aperture stop is opened too much — especially when objectives are used which are not of the highest degree of perfection
— the theoretical increase of the resolving power is offset by in-
sufficient contrast between the smallest resolved object structures and, possibly, also a detectable effect of the small residual aberrations in the passage of light through the entire optical
system.

The diameter of the area of the object plane, throughout
which coherence prevails when the NA of the illumination is reduced, increases continuously from a minimum, equal to the
limit of the resolving power of the objective, to a maximum
when coherence prevails throughout the entire field of view,
i.e., when the object is illuminated by unidirectional axial illumination with the aperture stop of the condenser closed to the
diameter of a "pin point".  The optical character of the image
changes under these conditions from optical fidelity of reproduction to complete lack of fidelity and greatly reduced resolving
power.

F.  <u>CONTROVERSIES</u>:  It is an unfortunate but undeniable fact
that after the Abbe theory became known, controversies arose
over the "correct" conditions for image formation.  One side
insisted that, regardless of the structures and optical character of the object, only unidirectional illumination (axial or
oblique) should be used.  The other side insisted that only the
optical characteristics of an image formed under conditions of
multidirectional illumination could be accepted as a true reproduction of the object.  We realize today that both uni- and multidirectional illumination have applications:  unidirectional illumination may be advantageous for revelation of the periodic structures of certain objects while multidirectional illumination is
much more acceptable for the illumination of objects with irregular structures.  These controversies were occasionally filled
with bitterness and animosity.  The practicing microscopists
who did not study publications in defense of either side and selected illumination conditions on the basis of analysis of the optical properties of the object and evaluation of the image quality
were fortunate to escape these controversies.*

G.  <u>DARKFIELD ILLUMINATION</u>:  There are many practically
colorless objects with refractive indices close to those of their
surroundings.  Some small living organisms, for instance, have
refractive indices of about 1.45 and when they are immersed in
water, they differ optically so little from their surroundings that
it is necessary to increase the contrast in the image (preferably
without loss of resolving power) for revelation of their structures.  Instead of achieving this by staining the object — which is
often impossible with living organisms — special illumination
methods can be used.  One of these is darkfield illumination.
The geometrical optical aspects of the performance of darkfield
condensers are described on pages 97 and 98 of Volume 14 and

-----

*It is regrettable but true that in some publications on mi-
croscopy, the term "Nelson-Abbe" or "critical-Abbe" illumination is to be found.  This is not only utterly wrong, but also
greatly misleading.  Abbe's theory is based on unidirectional
illumination and coherence in the object plane.  Nelsonian or
critical illumination produces the other extreme of complete incoherence in the object plane.

the lightpath through two types of darkfield condensers are shown
in Figure 43, Volume 14.

The upper glass surface of the <u>dry</u> darkfield condenser (on
the right side of Figure 43, Volume 14) is spherically concave
so that the cones of light emerging from that glass surface (at
all azimuths) proceed in radial directions without refraction towards the center of the sphere in the object plane. Because the
range of angles of inclination to the optical axis at which light
traverses the object plane is higher than the direction of
maximum inclination at which light can proceed to be collected by
the objective of given NA the field of view, with no object in the
lightpath, is completely dark.

The ranges of NA of two models of dry darkfield condensers
and the respective angles of inclination to the optical axis as well
as the respective ranges of the numerical apertures of the objective which are to be used with these darkfield condensers are
listed in Table III.

## Table III

### Angular apertures of two darkfield
### condensers and corresponding objectives

| Darkfield | Darkfield condenser | | | | Objective | |
| condenser | Max. NA | Max. angle | Min. NA | Min. angle | Max. NA | Max. angle |
|---|---|---|---|---|---|---|
| Model A | 0.95 | 71° 48' | 0.80 | 53° 40' | 0.75 | 48° 12' |
| Model B | 0.85 | 58° 13' | 0.70 | 44° 25' | 0.6 | 36° 52' |

The angular range between the highest NA of the objective
and the lowest NA of the darkfield illumination is required to provide darkfield illumination throughout a field of finite diameter.

A specimen in the object plane is made up of optical discon-
tinuities, due to reflection, diffraction and/or refractive index
difference. These optical interfaces cause changes in the direction of light so that it passes into the objective and, thus, object
details appear bright against the dark background.

When the object structures have smooth metallic surfaces, e.g., a net of very thin metallic wires, the changes of direction of light are caused by reflection.  When the refractive index of the object differs from that of the surroundings, the light is refracted by the structures of the object.  In each case, some of the light passes through the objective.  In both cases, light is also diffracted by the structures of the object.  Whereas the reflected and refracted light proceeds within relatively small angular ranges, the diffracted light passes through the entire range of the NA of the objective.  Under these conditions, the resolving power of the objective is equal to that for self-luminous objects, but the optical character of the image is not that of optical fidelity of reproduction.  It is the very purpose of darkfield illumination to increase contrast in the image and because of the reversal of the light conditions (dark background, bright object structures) the reproduction of the object is greatly improved. The difference between the bright object and the dark background can be increased by using light sources of higher intrinsic intensity.

Under conditions of darkfield illumination, very small particles can be seen (in collodial suspensions) even when their diameters are considerably smaller than the limit of the resolving power of the objective used.  These particles have ULTRAMICROSCOPICAL dimensions.  As the diameter of the particles decreases from a maximum size equal to the limit of resolving power of the objective to much smaller ultramicroscopic dimensions, the light diffracted by the particle passes through the objective and it becomes visible as a bright diffraction disc, the diameter of which does not decrease; only its light intensity decreases.  Each particle is still visible as a small diffraction disc of finite diameter as long as the lateral distance between adjacent particles is greater than the limit of the resolving power of the objective.

Incidentally, a similar effect can be observed at night on the cloudless sky.  Each star is seen within such a small viewing angle that it can be interpreted as a "point-shaped" object.  Each star is visible as a diffraction disc of finite, though very small, diameter.  At those parts of the sky where the lateral distance between "adjacent" stars is smaller than the limit of resolving power of the human eye, their images overlap and the area appears "cloudy". It is called "the milky way".

The greater the intensity of the light with which ultramicroscopic particles are illuminated and the greater the optical differences between the particle and the surroundings, the smaller is the smallest size of ultramicroscopic particles which can be seen with darkfield illumination. Suspended particles of colloidal silver and gold etc. can be seen even when their diameters are smaller than about 40 nm (compared to about 200 nm which is the limit of resolving power of oil immersion objectives of highest NA, i.e., 1.40).

The lightpath through a darkfield condenser for oil immersion is shown on the left side of Figure 43, Volume 14, page 97. These condensers are also immersion systems. The upper surface of the condenser is plane and perpendicular to the optical axis. The light proceeds through the upper portion of the condenser at angles of inclination to the top surface much higher than the critical angle of total reflection for passage of light from glass to air (Volume 14, pages 9-10). When the space between the upper surface of the condenser and the lower surface of the object slide is filled with oil of the same refractive index as the glass, light passes unrefracted through the object slide.

When the object is immersed in water, there is a limit of the highest angle of inclination to the optical axis at which light can proceed from the object slide (n = 1.515) into the water (n = 1.336) in which the object is immersed.

The maximum numerical aperture for passage of light from glass to water can be calculated as follows:

$$NA_{ill} = 1.515 \times \sin i = 1.336 \times \sin 90° \text{ and since } \sin 90° = 1$$

$$NA_{ill} = 1.336$$

Although darkfield condensers for oil immersions are listed with an upper limit of NA of 1.40, the light contributing to the illumination of objects immersed in water must be lower than 1.336, reducing the effective upper limit of darkfield illumination. For objects immersed in liquids of higher refractive index, the effective upper limit of darkfield illumination can be increased to the maximum of 1.40.

Oil immersion objectives are listed with NA up to 1.40. For darkfield illumination the NA of the objective must be reduced to provide an adequate angular range of the NA of darkfield illumination, even for objects immersed in water. This reduction can be achieved by installing in the optical system of the objective an iris diaphram acting as an aperture stop. For darkfield illumination, the iris diaphram must be partially closed so that the NA of the objective is lower than the low limit of darkfield illumination. For immersion darkfield condensers, this limit is generally about 1.20. To create darkfield illumination throughout the field of view of an oil immersion objective, its NA must be reduced to a value substantially lower than 1.20. One of the reasons for this reduction is that the iris diaphram cannot be placed in the plane for correct performance as an aperture stop. This correct plane is the back focal plane of the objective (Volume 14, page 42). In oil immersion systems of complex design, the back focal plane is somewhere within the lens system, actually within one of the lens components where it is impossible to install the iris diaphram. It is, therefore, placed as close to that plane as practically possible. It does not perform as an aperture stop with theoretical perfection. When a darkfield condenser with a low NA limit of 1.20 is used in combination with an oil immersion objective, the NA of which is reduced to only 1.15, observation of the image will disclose that only within a small central area of the field of view does the illumination approach darkfield; however throughout the field of view the illumination conditions are completely unacceptable. When the NA of the objective is reduced to about 1.00 or less, satisfactory darkfield illumination is obtained.

This range of unused NA of the objective, which has also been called the "lost" range of NA (between 1.20 and 1.00) reduces the resolving power of the objective as well as the intensity of the light in the image. It is possible to reduce this range of lost NA. Objectives have been produced, especially designed for darkfield illumination, with a maximum NA close to the lower limit of the NA of the darkfield condenser. They do not have an iris diaphram, but the mounts of the individual lens components are made of such diameters that one of them performs as an aperture stop with optimum efficiency. Leitz produced an oil immersion of high magnification with an NA of 1.15 to be used with a darkfield condenser of NA 1.20. The images formed by this objective

with darkfield illumination are noticeably brighter and have
higher resolving power. It seems that this objective is no longer
available. Possibly, the convenience of an iris diaphram, which
can always be opened to the optimum NA was considered more
desirable than an objective with greater resolving power for
darkfield illumination, but which, due to a lower maximum NA,
has lower resolving power for multidirectional brightfield illumination.

The slight loss of resolving power with darkfield illumination
assumed even less importance when illumination methods became available which made increase of contrast in the image
possible without loss of the full NA of the objective (phase contrast, interference contrast etc.). There are still numerous
applications of darkfield illumination where highest resolving
power is not essential.

A revival of interest in darkfield illumination has occurred
because of its advantages for fluorescence microscopy. The
complete <u>separation</u> of the NA range for the <u>illumination</u> of the
object with light of <u>short</u> wavelengths for <u>excitation</u> of <u>fluores-
cence</u> from the image-forming range for light of <u>longer</u> wavelengths <u>emitted</u> by fluorescence is exceptionally favorable for
this field of microscopy. Further details can be found in Volume
30.

Finally, there is the application of darkfield illumination to
"dispersion staining", a field of rapidly increasing importance;
it is described in Volume 35. To provide darkfield illumination
at low magnification, a special type of dry objective is equipped
with a "center" stop in its back focal plane so that light cannot
pass through the central area of the objective. By closing the
aperture stop of the brightfield condenser until its image in the
back focal plane of the objective is very slightly smaller than
the diameter of the center stop, darkfield illumination can be
produced and the outer regions of the lens system of the objective are used for image formation.

H. <u>VERTICAL ILLUMINATION</u>: An optical system for illumination of opaque objects with light of essentially vertical incidence,
for formation of images by light <u>reflected</u> by the object, is described in Volume 14, pages 64-65. The lightpath through the
vertical illuminator from the light source through components of

the illumination system, the 45° inclined reflecting glass plate and through the objective to the object, back again through the objective to the plane of the intermediate image, is shown in Figure 30 (Volume 14). At the lower surface of the glass plate (and also at its upper surface) partial reflection of the light occurs. The intensity of the reflected light is quite low compared to the transmitted part.

The reflected part proceeds through the lens components of the objective, which now performs as a condenser. Additional partial reflections occur at each entrance and exit surface of each lens component of the objective. The light passing through the central areas of the lens components proceeds at almost vertical incidence, and the percentage of the reflected part is relatively low, about 4% at each reflection. This light, however, proceeds after reflection in directions close to those near the optical axis. Light from these partial reflections reaches the image plane and the cumulative effect of all partial reflections is a considerable decrease of contrast in the image.

Light passing through the _outer_ (ring-shaped) areas of the lens components also undergoes partial reflection but, due to curvature of the lens components, this light, after reflection, proceeds in directions of higher inclination to the optical axis and is absorbed by the inside walls of the tube and by glare-reducing, ring-shaped diaphrams in the lightpath of the image plane.

Contrast reduction throughout the image, caused by partial reflections, is so great that, for instance, the images of objects with relatively low and diffuse reflection (paper, wood, textiles etc.) are practically worthless. In the passage of light from the objective to the image plane, when the objective performs as an image-forming system, further reductions of the light intensity occur because of partial reflections at the curved surfaces of the lens components.

I. <u>THIN FILM COATINGS</u>: Fortunately, it is possible to <u>improve</u> the ratio of intensities of the parts of the light partially reflected and transmitted by the glass plate. It is also possible to <u>reduce</u> the percentages of all partial reflections by the lens components of the objective.

1.  <u>Increasing the intensity of the partially reflected light</u>:
The intensity of the light, partially reflected at the surface of an
optically denser medium (proceeding from air into that medium)
depends on the angle of incidence and the refractive index of the
denser medium (Volume 14, page 15).  For incidence from air
into glass (n = 1.515) at an angle of 45° the percentage of partially reflected light is about 12-15%.  An increase of the percentage
of the reflected component can be achieved by coating the surface
with a material of substantially <u>higher</u> refractive index.

The same result, as far as the reflected component is concerned, can be achieved by using a material which is opaque at
greater thicknesses for coating the glass surface with a very thin
semi-transparent layer.  It is possible to control the extremely
thin layer of a deposit of silver (or other metals) so accurately
that any desired intensity ratio of the reflected and refracted components can be achieved.  Therefore, it is possible to improve the
ratio for optimum conditions of illumination and image formation.
When a metallic coating is used, the sum of the intensities of the
reflected and refracted components falls short of the intensity of
the light before reflection.  This is due to the <u>absorption</u> of light
in the silver layer.  When a transparent material of high refractive index is used for the coating, the sum of the intensities of
the two components is only slightly lower than the intensity before
reflection.

In either case, the result of the coating is an improved ratio
of the intensities of the two components which is beneficial for
image formation by reflected light.

2.  <u>Decreasing the intensity of the partially reflected light</u>:
The intensity of partially reflected light at the surface of a lens
component of the objective (or any other glass surface) can be
reduced by coating this surface with a <u>transparent</u> material having accurately determined refractive index (<u>lower</u> than that of the
glass) and a definite thickness.  The cause of this reduction of the
intensity can be explained with the aid of Figure 33.  On the left
side, the reflection from an <u>uncoated</u> surface is shown.  Light,
proceeding in the direction 1 at almost vertical incidence, is
partly reflected and partly refracted.  When the refractive index
of the glass plate is 1.515, the percentage of the reflected component is about 4%.

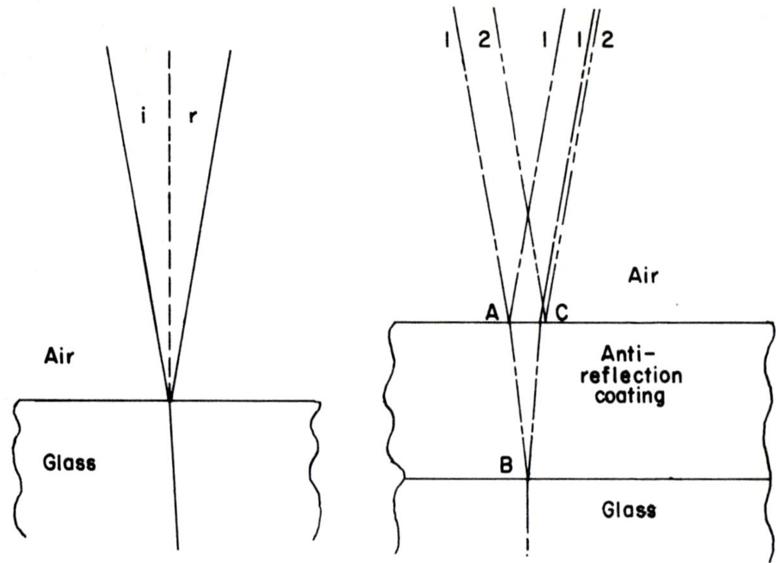


Figure 33.  The effect of an antireflection coating

On the right side, partial reflection from a <u>coated</u> glass
plate is shown.  Ray No. 1 is partially reflected at the <u>upper</u> surface of the coating which has a refractive index, <u>lower</u> than that
of the glass plate.  The percentage of its intensity is lower than
4%.  The refracted component (of much higher intensity) passes
through the thin layer from point A to B where another partial
reflection occurs.  This component proceeds through the thin layer from B to C where it emerges into air.  Ray No. 2, originating in the same point at infinity as ray 1, proceeds parallel to
ray 1 and arrives at point C where it also undergoes partial reflection.  There is coherence between the partially reflected
components of rays 1 and 2 proceeding from point C beeause they
originate in the same point.  These rays, therefore, interfere
with a pathlength difference of the magnitude of AB + BC multiplied by the refractive index of the layer.  By correct selection
of the thickness of the layer and its refractive index, the two rays
interfere with a pathlength difference of one-half wavelength but
equal amplitudes and they cancel each other.  Since white light is

composed of wavelengths from 400 to 700 nm, complete elimination of the partial reflections cannot be achieved. Practical experiments, however, have resulted in developing a technique of providing the lens surfaces with an ANTIREFLECTION COATING which reduces the detrimental effect of these reflections considerably. The components of practically all lenses of optical systems, _e.g._, objectives, oculars and condensers of microscopes, are antireflection coated. Coated surfaces show a bluish tint when observed in reflected light. The effect of these coatings is particularly beneficial for image formation with vertical illumination by reflected light because light passes through the objective, not only on its way to the object plane (to illuminate the object) but also a second time for formation of the image.

Vertical illumination is used extensively for polished and etched surfaces of metals with relief structures. In spite of the antireflection coating, this illumination is still not satisfactory for diffusely reflecting objects. Figure 34 is a photomicrograph of a metal surface, polished and etched. The relief structures are reproduced with good contrast. Figure 35 is a photomicrograph of a piece of white paper with black printing on it. The contrast is so low that the black print appears gray. Figure 36 is a photomicrograph of the same piece of paper by oblique darkfield illumination. The contrast is greatly improved.

## J.  "EPI" ILLUMINATION WITH HIGH ANGLES OF INCIDENCE:

The illumination method used for the photomicrograph of Figure 36 is similar to darkfield illumination inasmuch as the object is illuminated with light of a range of higher NA than that of the objective. The lightpath through two top light illuminators is shown in Figure 37 and 38.

Light from the source, after having passed through the collector, enters either the "Epi" or ultropak illuminator in a horizontal direction. It is reflected by a plane, elliptically-shaped

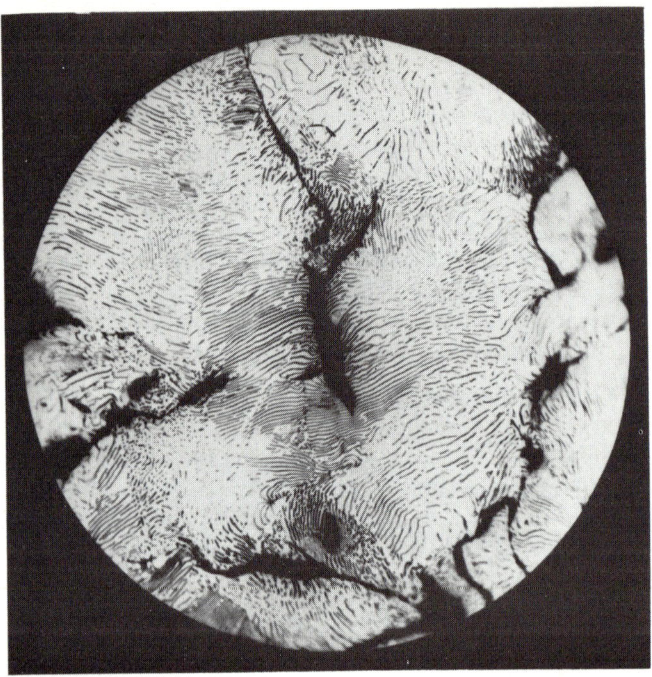

Figure 34.  Photomicrograph of perlitic steel, polished and etched, showing relief structures; vertical illumination.

annular mirror into a vertically downward direction.  It proceeds outside of the objective mount, so to speak, as a "hollow cylinder" of light and is reflected once more by a curved mirror*, correctly placed with respect to the lens system of the objective so that

---

*In the Leitz Ultropak, one of the first systems for darkfield illumination by reflected light, the light passes through a condenser system as those lenses have central holes of suffciently large diameter for the cylindrical mount of the objective to pass through them (Figure 38).

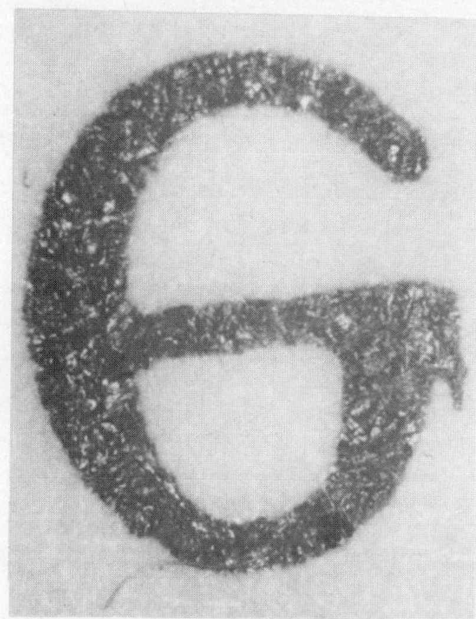

Figure 35.  Black print on matte paper; vertical illumination.

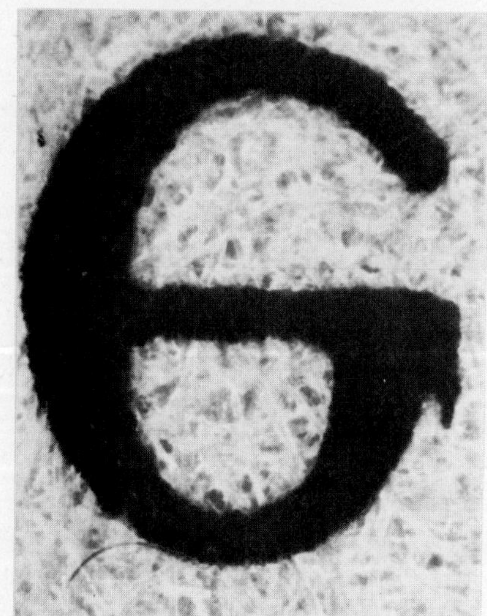

Figure 36.  Same as Figure 35 but with oblique "Epi" illumination (reflected dark-field).

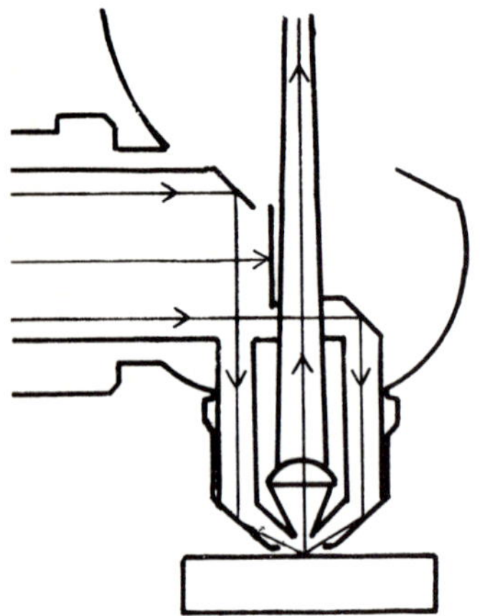

Figure 37. Zeiss "Epi" condenser for reflected darkfield illumination with a mirror condenser.

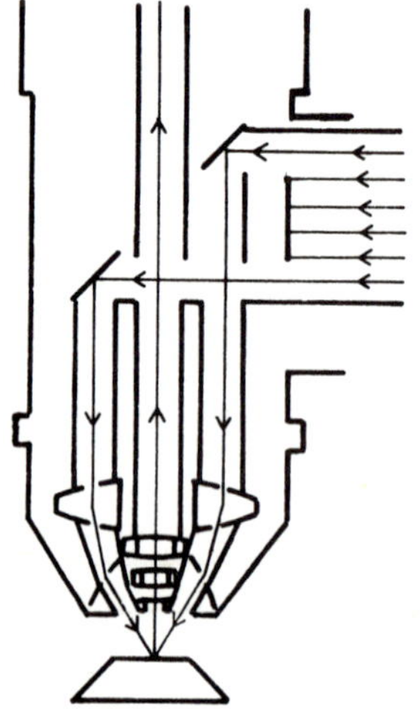

Figure 38. Leitz "Ultopak" illuminator for reflected darkfield illumination with a lens condenser.

it concentrates light on the object with higher numerical aperture than that accepted by the objective.  The passage of illuminating light rays through the lens components of the objective is thus avoided.  The light is diffusely reflected by an object like paper etc. and passes through the objective and a hole of adequate diameter in the elliptical, inclined mirror to form the image in the lower focal plane of the ocular.

When this illumination method is used for image formation of metallic objects with polished and etched surfaces and relief structures, the optical character of the image is similar to that produced by darkfield illumination in transmitted light.  When two photomicrographs are taken of the same object, one with vertical illumination, the other with Epi illumination with high angles of incidence, there is generally a reversal of contrast similar to that between a negative and a positive.  However, there are exceptions which may be of practical significance.  One of them is shown in Figure 39.

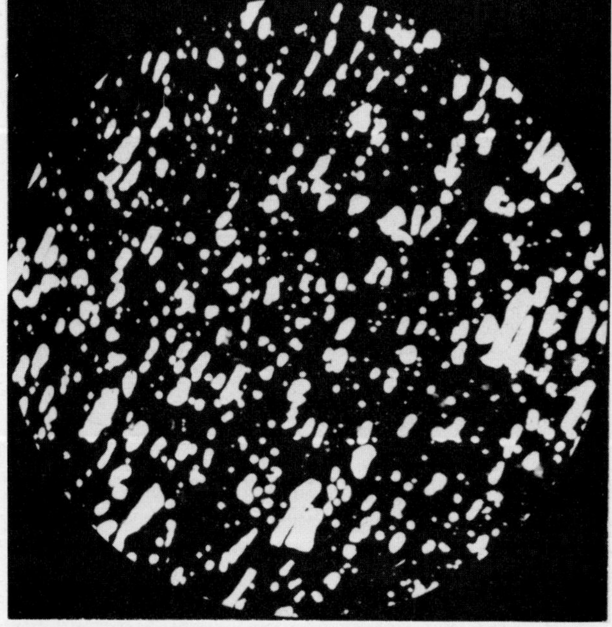

Figure 39.  Ferrite (oxidized) with carbide inclusions shown by vertical illumination.

The first photomicrograph was taken with vertical illumination. The object is a piece of steel, polished and etched to show the elongated "islands" of carbide inclusions against the background of ferrite. The reason why the background appears dark is that the object was heated so that the ferrite was oxidized and its reflectivity was greatly reduced. Therefore, it appears dark in the photomicrograph which was taken with high contrast film. The reflectivity of the plane surfaces of the carbide inclusions are undiminished and they appear white.

In Figure 40 "indirect" (darkfield) illumination was used.

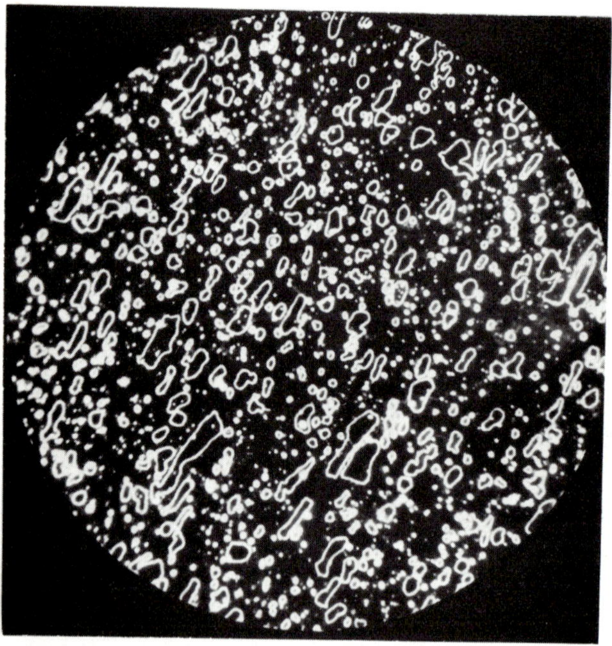

Figure 40. Same object as in Figure 39 but with reflected darkfield illumination.

Since the surfaces of the ferrite as well as the carbide inclusions are plane and smooth they reflect the light into directions of as high an inclination to the optical axis as the illumination. They appear equally dark in the picture. But the slopes of the elevated inclusions are at high angles towards the object plane and reflect the light into the objective. They appear bright in the picture. The picture of Figure 40 also shows numerous white "dots", caused by ultramicroscopically small inclusions which could not be seen with vertical illumination.

Since opaque objects are generally not covered by cover slips, these (metallographic) objectives are not corrected for cover slip. Objectives used for image formation by transmitted light are cover slip corrected (Volume 14, pages 71-73).

K. RESOLVING POWER AND IMAGE CHARACTER: There is an analogy between the optical conditions of image formation by transmitted and reflected light systems; both are based on Köhler illumination. They both have aperture and field stops of variable diameters and the resolving power and the optical character of the image vary with the NA of the illumination and the objective as described.

The analogy also prevails for darkfield illumination and Epi illumination at high angles of obliquity. Incidentally, the actual magnitudes of the smallest resolved object structures of self-luminous and nonself-luminous objects, according to the equations quoted in this book, are so close to each other that it has become common practice to use only one equation:

$$d = \lambda/2\,NA \qquad\qquad \text{Equation (13)}$$

This equation is, however, based on conditions of extremely oblique, unidirectional illumination for which it is fully valid. When objectives of simple design (achromats) are used with monochromatic light, it should be kept in mind that for multi-directional illumination (especially with reduced NA) the resolving power does not fully reach the value expressed in this equation.

## RESOLVING POWER
## AND LIMIT OF USEFUL MAGNIFICATION

A.  WHAT IS THE LIMIT OF USEFUL MAGNIFICATION ?  Every image-forming optical system has a limited resolving power, including the human eye.  When the smallest object detail resolved in the image formed by an objective of given NA has been magnified by the ocular so much that in the visually observed image it has a magnitude equal to the limit of the resolving power of the human eye, any additional magnification — for instance, by using an ocular of higher magnification — does not result in the revelation of smaller object detail.  If the magnification of the visually observed image is lower, the smallest object detail resolved by the optical system of the microscope cannot be seen.

The total magnification of the image, formed by the optical system of the microscope which must prevail so that the smallest resolved detail is equal to the limit of resolving power of the observer's eye, is called the LIMIT OF USEFUL MAGNIFICATION.  Any further increase of the magnification results in EMPTY MAGNIFICATION.

B.  THE RESOLVING POWER OF THE HUMAN EYE:  The resolving power of the human eye depends not only on the physical optical performance of its lens system, but also on biological factors; for instance, the diameter (about 5 μm) of the nerve cells on the retina on which light impinges and from which the stimuli caused by variations of the light intensity are transmitted to the brain where they are integrated into the image which we see.

The physical optical factors on which the resolving power of the eye depends are: the wavelength of the light forming the image, the focal length of the lens system of the eye and its effective diameter.  The focal length varies, depending upon the distance between the object and the eye.  When looking through the microscope, the eye, in its state of greatest relaxation, is accommodated to an object distance of infinity and under these conditions, its focal length is about 15 mm.

Also the effective diameter of the lens varies, because the iris diaphram which controls the diameter of the pupil of the eye contracts when light of high intensity enters it and expands under conditions of reduced light intensity.  When looking through the microscope, the diameter of the pupil of the observer's eye should be greater than that of the exit pupil of the microscope (Volume 14, page 48).  When the light intensity of the image is adjusted for "comfortable" conditions of visual observation, the diameter of the pupil is about 2 mm.  For image formation of an object at infinity — considering only the physical optical aspects of the performance of the eye — the limit of the resolving power can be calculated by using Equation (4) (page 23) with certain modifications.  For image formation by the eye, the distance from the aperture stop to the plane of the image is its focal length, i.e., the distance from its iris diaphram to the retina.  The radius R of the lens is 1 mm.  The wavelength is reduced from $\lambda$ to $\lambda/n$ where n is the refractive index of the fluid in the eye, 1.336.  Since the radius R  is small in comparison with the focal  length f of the eye, $\sin \beta$ can be replaced by $\tan \beta$, ($\tan \beta = R/f$).

When these values have been inserted into Equation (4) (page 23), the magnitude of r, the radius of the diffraction disc, which is the limit of the resolving power of the eye, can be determined as follows:

$$r = 1.22\,f\lambda/2\,nR = 1.22(22,800)(0.55)/2(1.336)(1000) = 5 \;\mu m$$

($\lambda$, R and f must be expressed in micrometers.)

This value of 5 $\mu$m is the distance in the image, i.e., on the retina of the eye.  The distance in the object corresponding to 5 $\mu$m in the image is:

$$5 \text{ x } \frac{\text{object distance}}{\text{image distance}} = 5 \text{ x } \frac{25,4000}{15,000} = 84 \;\mu m$$

The limit of the resolving power of the eye is generally expressed as the minimum viewing angle which must prevail between two adjacent object points so that in the image, the distance between the centers of the diffraction discs is equal to the radius of a diffraction disc.  According to simple calculations, this angle is about one minute of arch which corresponds to about 73 $\mu$m.

In practice, a wider minimum viewing angle has been accepted as the limit of the resolving power of the human eye.  In the first place, the calculated value is based on the assumption that the image-forming lens system is of theoretical perfection, free from all aberrations.  The lens system of the human eye does not meet this requirement. In the second place, the biological factors which influence its resolving power must be considered. The diameters of the nerve cells on the retina which transmit the light stimuli to the brain are about equal to the limit of the resolving power of the eye but a small amount of the light from each diffraction disc falls within the area of the surrounding nerve cells.

Altogether, the conditions of image formation by the eye are very complex and calculations based only on the physical optical aspects of image formation yield values for the resolving power which are never fully reached in reality especially under conditions of prolonged viewing. An approximate range of the magnitudes of the viewing angles which correspond to the limit of the resolving power of the human eye is from 2-4 minutes of arc.

For the following calculations of the approximate limit of useful magnification for visual observation, a viewing angle of 3 min has been used.

In Volume 14, page 34 and Equation 6b, the magnification of the ocular has been defined as the ratio of the viewing angle within which an object of given length is seen through the ocular to that within which the same object appears when viewed by the unaided eye from a distance of 250 mm.

Two adjacent object points, when viewed by the unaided eye from a distance of 250 mm, which appear within a viewing angle of 3 min are separated by a lateral distance D of about 0.22 mm because:

$$\tan 3 = \frac{D}{250}; \qquad D = 250 \times \tan 3 = 0.218 \text{ mm } (218 \text{ } \mu m)$$

This distance has been accepted as the approximate limit of the resolving power of the human eye, not only for microscopy, but also for looking at any object from a distance of 250 mm (10 in.).

Based on these figures, the half-tone illustrations in books and journals are composed of individual "dots", separated by a distance which is slightly smaller than the limit of the resolving power of the eye. Half-tone screens having 40-175 lines per inch are used in the printing industry for reproduction and printing of pictures. The resulting dots on the printed page are a convenient test of resolving power of the eye (Table IV).

## Table IV

### Spacing of half-tone dots

| Use | Dots/in. | Spacing ($\mu$m) |
|---|---|---|
| Newspapers | 40–85 | 640–300 |
| Machine-finished papers, bonds | 100 | 250 |
| Dull-coated stock, high-grade bonds | 120 | 210 |
| Coated stock | 133 | 190 |
| Fine books and catalogs | 150 | 170 |
| Highest grade of coated stock | 175 | 145 |

The number of dots/in. is easily counted with a ruler and 8X–10X hand magnifier.

In practice, not only books but also photomicrographs are generally held at distances greater than 10 in. for convenient viewing. The longer the viewing distance, the greater is the actual magnitude of the finest object detail which can be resolved by the human eye appearing within the viewing angle of 3 min. On the other hand, if your unaided eyes can accommodate an image only 5 in. away the magnification is then 2X and you should be able to resolve two points half as far apart as your limit at the 10 in. viewing distance.

C. <u>THE LIMIT OF USEFUL MAGNIFICATION IN VISUAL OB-SERVATION</u>: When the smallest object detail of the magnitude, d, resolved in the image formed by an objective of given NA, has been magnified by the combined performance of the objective $M_{obj}$ and the ocular $M_{oc}$ so that its image has the same magnitude, D, as that of the smallest detail which the eye can resolve, <u>the limit of useful magnification</u> has been reached.

$$M_{obj} \times M_{oc} = D/d; \quad d = \lambda/2\,NA; \quad D = 250 \times \tan 3;$$

Therefore: $M_{obj} \times M_{oc} = 250{,}000 \times \tan 3' \times 2 \times NA/\lambda =$

(approx.) 800/NA (250 mm = 250,000 $\mu$m, $\lambda = 0.55\ \mu$m)

For practical purposes and taking into consideration the variations of the eyesight of different observers, it may be assumed that the limit of useful magnification is approximately 500–1000/NA of the objective, forming the intermediate image.

The values given in Table V are based on an average value of this limit of 750 NA.

## Table V

### Ocular magnifications required for a selection of objectives

| Objective | $M_{obj}$ | NA | 750 NA | Approximate $M_{oc}$ |
|---|---|---|---|---|
| Planachromat | 2.5X | 0.08 | 60X | 25X |
| Planapochromat | 4X | 0.16 | 120X | 30X |
| Achromat | 10X | 0.22 | 165X | 16X |
| Planapochromat | 10X | 0.32 | 240X | 25X |
| Planapochromat | 25X | 0.65 | 487X | 20X |
| Achromat | 40X | 0.65 | 487X | 12.5X |
| Planapochromat | 100X | 1.32 | 990X | 10X |

The ocular magnification required to reach the limit of useful magnification varies within relatively wide limits. In general, it is higher for objectives of low magnifications. Also, for any given objective magnification, it is higher for planapochromats because of their relatively higher numerical apertures.

In practice, magnifications deviating considerably from the limit of useful magnification are often used. For instance, at low magnifications, for scanning the "topography" of a histological, stained section, it is desirable to increase the diameter of the object area within the field of view as much as possible. An objective of 2.5X magnification may be combined with a wide angle ocular of 10X magnification, revealing an area of the object plane of about 7.2 mm within the field of view. If the same objective is used with an ocular to form the final image at the limit of useful magnification, it would have to be combined with an ocular of 25X magnification and the diameter of the object area revealed within the field of view would be about 2.5 mm. of the smallest resolved object detail.

On the other hand, it is often advantageous at high magnifications to exceed the limit of useful magnification. For counting of small particles or for observation of small organisms (<u>e.g.</u>, bacteria), it is often more comfortable to view the image at higher magnification although its sharpness may not be at an optimum. For special purposes, magnifications exceeding the limit of useful magnification by as much as 100% and more are often used.

In order to avoid frequent changes of oculars when using objectives of low to high magnifications and in order to change the total magnification to suit the special conditions of observation, it is possible to interpose between objective and ocular an intermediate optical unit, (<u>e.g.</u>, Zeiss OPTOVAR; Volume 14, page 89) with telescopic systems of several intermediate magnifications. The total magnification can be changed by selecting one of the components of the intermediate systems without changing the ocular and without interrupting observation of the object.

D. <u>LIMIT OF USEFUL MAGNIFICATION IN PHOTOMICROGRAPHY</u>: In the image <u>recorded on the film</u>, the magnitude of the smallest resolved object detail also depends upon the capacity of the emulsion to <u>record</u> the detail. This capacity is called PHOTOGRAPHIC RESOLVING POWER. There is a difference between the PHYSICAL RESOLVING POWER of a lens system and the photographic resolving power of the film.

The <u>physical resolving power</u> of a <u>photographic lens system</u> forming the image when the object is at <u>infinity</u> is considerably higher than the photographic resolving power of the films generally used. It also depends, in addition to the wavelength of the light which forms the image, on the focal length and effective diameter of the lens system. The ratio of these two factors is called the F-VALUE of the lens system.

The diameter of the diffraction disc, the image of a single point at infinity, formed by a lens with an F-value of 3.5 is about 5 μm. An area of this diameter is illuminated on the <u>surface</u> of the light-sensitive <u>emulsion</u> which has a <u>finite thickness</u> and is <u>turbid</u>. As light penetrates through the emulsion, it is <u>scattered</u>. Silver bromide grains within an area of larger diameter than that of the diffraction disc are exposed to light. After development, the silver grains throughout the thickness of the emulsion decrease the photographic resolving power of the film.

Black and white films, 40-50 years ago, had much lower resolving power than the lenses of miniature cameras which started to appear at that time. The average film of medium sensitivity could resolve about 30-40 lines per mm, compared to the photographic resolving power of the lens which exceeded 400 lines per mm. But even those films could record detail much finer than the human eye (4-5 lines per mm) at 10 in. That is the reason why the small negatives could be enlarged to much larger sizes and why miniature cameras were so successful.

Remarkable progress has been made since then, not only in the production of films with higher photographic resolving power, but also with finer "grain". Incidentally, the smaller grain size does not necessarily cause a higher photographic resolving power. On the contrary, fine grain films of some 40 years ago had a somewhat reduced resolving power because the greater number of smaller grains caused a greater <u>scattering</u> effect of the light passing through the emulsion.

Films for black and white photography, currently produced for miniature cameras (35 mm) have a photographic resolving power of about 150 lines per mm and finer. Even so-called fast films have a remarkably fine grain and high resolving power.

In the image formed by the optical system of the microscope for photomicrography at a distance of 10 in. at the limit of useful magnification, about 4-5 lines per mm are resolved. If a photomicrograph is taken with an attachment camera on 35 mm film and if the auxiliary lens of the camera reduces the magnification to 1/3 (Volume 14, page 35), the limit of the resolving power of the optical stystem is still equal to only about 15 lines per mm. The photographic resolving power of the film is still about 10 times as high.

Because of this fact, attachment cameras using 35 mm film can be used to great advantage in photomicrography, especially in those cases where reduction of the exposure time is of practical advantage. The image recorded on the 35 mm film can be enlarged to a print size of 8x10 in. and all of the object detail which the image-forming system of the microscope could resolve is reproduced in the enlargement.

Also color films of current production have remarkably high

photographic resolving power.  Literature is available from the film manufacturers giving information about the photographic resolving power of these as well as black and white films.

E.  <u>LIMIT OF USEFUL MAGNIFICATION IN PHOTOMACROG-RAPHY</u>:  The field of photomacrography can be defined as photography without a microscope, from natural size to relatively low ranges of magnifications.  Special MACRO lens systems are available for this purpose.  They are highly corrected photographic lens systems, corrected especially for image formation at relatively short image distances, varying from a minimum to a maximum, corresponding to the range of variable bellows extensions of cameras for photomicrography (about 10-25 in.).  Macro lens systems are available with focal lengths from about 16 mm to about 120 mm.

Certain figures are engraved on the mounts of <u>micro</u> objectives.  They specify the magnification (for the intermediate image) at which the quality of the image is at an optimum and also the NA of the objective.  The user of these objectives can determine the magnification of the ocular required to reach the limit of useful magnification of the final image and also the magnitude of the smallest object detail which can be resolved by the selected objective.  For objectives of high NA, the specified magnification must be maintained within narrow tolerances.  A deviation of as little as 10 mm from that image distance at which the specified magnification prevails causes detectable decrease of image quality.  That is why micro objectives are attached to tubes of fixed mechanical tubelength.

Macro lens systems have much lower ranges of NA and the tolerances for deviations from the image distances at which the image quality is at an optimum are much greater.  The NA of a macro lens system does not have a constant value.  It varies with the object distance.  This is shown in Figure 41.

In the diagram on the left side, the magnification is 1X.  The image distance is equal to the <u>object</u> distance, 2f (Volume 14, page 27).  In the diagram on the right side, the magnification is 5X and the object distance has decreased to 6f/5.  According to calculations, based on an equation which will be developed in the further course of this chapter, the NA of the macro lens system has increased from 0.071 to 0.113 (assuming that the F-value of

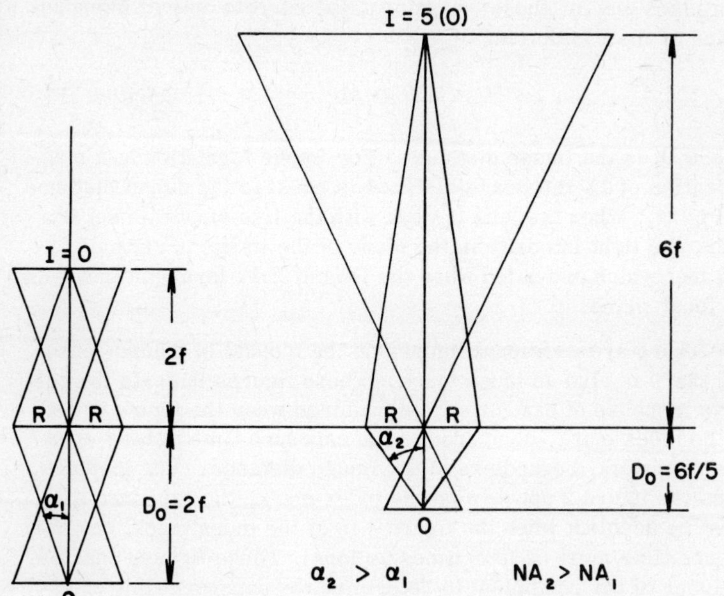


Figure 41.  The NA of a macro lens system increases with
magnification.

the macro lens system is 3.5).

Figures are also engraved on the mounts of <u>macro lenses</u>.
One of them is the F-value which indicates the relative light
transmitting capacity of a lens for image formation when the object is at infinity and the image is formed in the focal plane of the
lens.  Under these conditions:

$$F = f/d \qquad \text{Equation (14)}$$

where F is the F-value, f is the focal length of the lens and d is
its effective diameter.  The light transmitting capacity of a lens
decreases with the square of its F-value.  For instance, the light
intensity in the focal plane of a lens when its iris diaphram has
been stopped down to F:9 is 1/4 of that when the iris diaphram has
been opened to F:4.5.

When the object is at a finite distance, the image distance increases and for these conditions, the correct optical interpretation of the F-value is:

$$F = D_i/d$$ 
Equation (15)

where $D_i$ is the image distance. For image formation at a magnification of 1X the image distance is equal to the object distance and is 2f. When the lens is used with the iris diaphram set to F:22, the light intensity in the plane of the image is reduced to 1/4 that which prevailed when the lens at F:11 forms an image in its focal plane.

There are additional figures on the mounts of macro lenses and there is also an index mark. These figures indicate the relative increase of exposure time required when the figure is set to the index mark, compared to the exposure time at fully open iris diaphram, regardless of the image distance. For instance, when the figure 2 appears on the index mark, the exposure time must be doubled; when the figure 4 is at the index mark, the exposure time must be four times as long. These figures enable the user of the equipment to determine the exposure time at reduced effective diameter of the lens when the correct exposure time with a fully open iris diaphram is known. Quite often, the exposure time is determined at full lens aperture and the iris diaphram is subsequently closed to obtain the desired depth of field.

With the iris diaphram partly closed, the light intensity in the image plane may be too low for exact determination of the exposure time.

For determination of the limit of useful magnification of a macro lens it is necessary to find the geometrical relationship between the F-value of the lens and its NA for a selected magnification. According to Figure 41 the following relationships exist between the object distance $D_O$, the radius R of the lens and the angle $\alpha$ between the optical axis and the direction of the light of highest inclination to it which can be collected by the lens:

$$\sin \alpha = R/\sqrt{D_o^2 + R^2} = NA \quad [2R = d]$$

Since $F = f/2R$ and $R = f/2F$; since, furthermore $D_o = f(M+1)/M$

(Volume 14, Equation 4a)

$$NA = M/\sqrt{4F^2 (M+1)^2 + M^2} \qquad \text{Equation (16)}$$

Equation (16) is used for calculating the NA of a macro lens for given F-values and magnifications. Some of these values are listed in Table VI.

Values of NA for an F-value of 2.5 and magnification of less than 10X have been omitted because only the macro lenses of shortest focal lengths are available with this F-value and, even at the shortest bellows extensions, magnifications of less than 10X are not obtainable.

According to this tabulation, the NA of a macro lens of F = 2.5 used for image formation at 10X magnification has a NA of 0.178, and the limit of useful magnification is as high as 133X. The negative can be enlarged about 13X in order to magnify the resolved object detail so that it can be seen when the enlargement is viewed from a distance of 10 in.

The _highest_ magnification obtained with _macro_ lenses and a camera with bellows of variable extension is higher than the _lowest_ magnification obtained with a _micro objective_ of lowest magnification. At equal magnifications, the NA of the macro lens is higher and, therefore, also its limit of useful magnification. For instance, using a Zeiss Planachromat 2.5X (NA = 0.08) with an ocular of 10X magnification and a bellows extension of 16 in., the magnification of the image in the plane of the film is 40X. The limit of useful magnification for this objective is 60X (750x0.08 = 60). On the other hand, using a 16 mm Zeiss Luminar at full aperture (F = 2.5) and a bellows extension of about 26 in., the magnification is also 40X. The NA of the Luminar at the magnification is 0.191 and the limit of useful magnification is 143X.

Macro lenses are generally equipped with iris diaphram for reduction of their light transmitting capacity and increase of the depth of field. It is true that at reduced NA the resolving power

## TABLE VI

Numerical apertures of a macro lens for given
F-values and magnification

| Magnifi-cation | F-Values | | | | | | | |
| --- | --- | --- | --- | --- | --- | --- | --- | --- |
| | 2.5 | | 3.5 | | 4.5 | | 6.3 | |
| | NA | LUM* | NA | LUM* | NA | LUM* | NA | LUM* |
| 1X | | | 0.07 | 52X | 0.055 | 41X | 0.040 | 30X |
| 2X | | | 0.09 | 67X | 0.074 | 55X | 0.053 | 39X |
| 5X | | | 0.118 | 88X | 0.092 | 69X | 0.066 | 49X |
| 10X | 0.178 | 133X | 0.128 | 96X | 0.100 | 75X | 0.071 | 53X |
| 20X | 0.187 | 140X | 0.135 | 101X | 0.105 | 78X | 0.075 | 56X |
| 40X | 0.191 | 143X | 0.139 | 103X | 0.107 | 80X | 0.077 | 57X |

*LUM = Limit of useful magnification.

of the macro lens decreases, but there is still a relatively large
margin before the decreased resolving power becomes detectable
in the (unenlarged) print.

## USING THE MICROSCOPE TO THE LIMIT OF ITS CAPACITY

**A.   THE LIMITING CAPACITY OF THE LIGHT MICROSCOPE:**
With the introduction of apochromats by Professor Abbe in 1886, microscope objectives were improved so much that their resolving power was practically equal to the limit imposed by the physical nature of light.  Since then, only minor improvements in the corrections of the best objectives — except in regard to greater flatness of the field — have been made.  As far as the resolving power is concerned, the microscope can be used to perform to the limit of its capacity by following a systematic and simple procedure which will be outlined in detail in this chapter.

There is, however, another important aspect to the capacity of the microscope.  In the introduction to Volume 14, it was mentioned that any object, whether large or small, can be seen only if it differs, optically, sufficiently from its surroundings. When these differences are very small and cannot be increased by special methods of preparation of the object, illumination methods must be found to reveal the object detail in the image with adequate contrast and without loss of resolving power.

The most significant developments in microscopy during the past 40-50 years concern the creation of new illumination methods and equipment to achieve this.  An indication of the practical importance of some of these developments is the fact that Dr. Zernike, the creator of phase contrast illumination, was awarded the Nobel prize.

With these new illumination methods, object detail can be revealed in the image with good contrast — under some conditions with variable contrast — even when the object differs optically only slightly from the surroundings.  New methods of microscopy go beyond the revelation of morphological detail of the object and include qualitative as well as quantitative analysis of the optical properties of the objects, e.g., interference microscopy.

Because of these developments, the limits of the capacity of the microscope as an optical instrument have been widened considerably.  Furthermore, revelation of object detail, very much smaller than the limit of the resolving power of the light microscope has been achieved by the electron microscope.  The continued great practical value of the light microscope lies in its versatility.

Detailed descriptions of the optical principles and capabilities of other illumination methods are to be found in other volumes of this series. Phase microscopy is described in Volume 28, interference microscopy in Volume 29, fluorescence microscopy in Volume 30, microscopy in the ultraviolet in Volume 24, and in the infrared in Volume 25.

Microscopes of current production are designed for great versatility and for convenient interchangability of accessories for various illumination methods.

B. SELECTION OF EQUIPMENT: There are three major categories of microscopes:

1. Microscopes for image formation of translucent objects by transmitted light.

2. Microscopes for image formation of opaque objects by reflected light.

3. Microscopes of greatest versatility for image formation with transmitted as well as with reflected light. In this category there are microscopes with practically unlimited versatility which can be used not only for visual observation and photomicrography but also for computerized optical image analysis.

All of these microscopes have built-in illumination systems. In the first category, there are two basic types of stands: upright and inverted microscopes. The first type is used most frequently for biological, medical, mineralogical and chemical microscopy. Inverted microscopes — with the object stage above the objective — are designed for image formation of tissue cultures which are generally grown in flasks, petri dishes or similar containers. When one of these containers is placed on the object stage of an upright microscope, the working distance of most objectives is not long enough to focus the image of, say, a culture grown in a petri dish. Objectives with adequate working distances for formation of the image of the culture on the upper surface of the relatively thin bottom plate of the container are available, even with relatively high magnifications. Suitable condensers for various illumination methods are also available

with unusually long working distances. When the culture in its
container is placed on the object stage of an inverted microscope,
it can be illuminated by a suitably selected condenser.

One of the advantages of the built-in illumination system is
that the alignment of its optical components for optimum illumination conditions in the object plane is based on an inalienable
optical criterion: the formation of the image of the field stop in
that plane. As previously explained, the field stop is placed in
a plane of the illumination system where the intensity of the light
does not vary throughout an area at least as large as the fully
open field stop.

In former years, light sources with collectors and with or
without field stop were on separate stands. The alignment of the
illumination system was more difficult and uncertain. Furthermore, it could be disturbed by involuntary movements of the
light source of the microscope. The built-in illumination system,
once it has been aligned (by a simple and systematic procedure),
remains in alignment.

The _light source_ is generally an incandescent lamp, with concentrated filament for low voltage, in a precentered and prefo-
cused socket. A light source of small dimensions (with low heat
dissipation and current consumption) can be used because the collector forms its _magnified_ image in the plane of the aperture stop
of the condenser. Light sources of larger dimensions (and
greater heat dissipation because of higher wattage) would not increase the light intensity in the object plane. Their magnified
image would be larger than the diameter of the fully open aperture stop and would be prevented from reaching the object plane.

_Adjustable transformers_ (either stepwise or continuous) are
supplied to reduce the line voltage feeding the lamp and to vary
its intensity. The light intensity required for highest magnifications and binocular observation (as well as for photomicrography
and for other illumination methods) is much higher than that for
lower magnifications. There are some built-in illumination
systems which do not meet all of these requirements. It is advisable to find out if the light intensity at highest magnifications
and binocular vision is still satisfactory and if the illumination
system has a field stop. Its absence will be a distinct disadvan-
tage.

Reduction of the light intensity can be made either by lowering the voltage feeding the lamp or, preferably, by interposing a neutral density filter. A lower voltage for the lamp causes a change of its "color temperature". This is not particularly objectionable for visual observation when a blue (daylight) filter is interposed in the lightpath. It is also not objectionable for photomicrography with black and white film. When color film is used, the light source must be used at a specified voltage because its color temperature must be matched (or corrected with color-balancing filters) to the color sensitivity of the film.

The substage is generally permanently attached to the microscope stand. It must be equipped with a focusing device for the condenser to focus the image of the field stop in the plane of the object and a centering mount for the condenser to center its optical axis with that of the objective (when the image of the field stop is concentric with the periphery of the field of view).

The brightfield condenser generally consists of several components which can be used either as a complete unit or with some of the components removed from the lightpath in order to create correct illumination conditions in the object plane from lowest to highest magnifications. The removable components are either in swing-out mounts or (in aplanatic-achromatic condensers of highest state of correction) they must be removed by unscrewing them. This method is less convenient but preferable for maintenance of the centration of all components to one common optical axis, when the highest degree of accuracy is required for optimum performance.

Selection of the components of the condenser system for illumination of the object from lowest to highest magnifications varies for different manufacturers. The booklet accompanying each unit should be consulted about the recommended method to produce optimum illumination conditions at all magnifications.

In view of the availability of various illumination methods for transmitted light, the prospective user of a microscope should inquire from the manufacturer, whose product he intends to purchase, which of the accessories for these additional methods are available for interchangeable use on the selected microscope. This is not necessary, of course, if the microscope is to be used only for brightfield illumination in a single field of application.

Several object stages are available.  Any one of them can be selected by the prospective user and it will be attached by the manufacturer before delivery is made.  Plain square object stages with attachable or built-in mechanical stages (graduated or ungraduated) are used most frequently in biological  micros-copy.  Circular, rotating object stages with attachable or built-in mechanical stages are also available.  There are situations when a rotating stage is preferable or even essential, i.e., polarized light microscopy.

A rotating stage may even be preferred when the microscope is used for photomicrography since the image can be arranged for most favorable orientation within the rectangular film frame. These stages are equipped with centering devices so that the ro-tation axis of the stage can be aligned to coincide with the optical axis of the selected objective.

When several objectives are attached to a multiple nosepiece, it is practically impossible to achieve a state of perfection of cen-tration of all of them to one common optical axis within a very small tolerance.  Therefore, minor corrections can be made with the centering device of the rotating stage to center the image for rotation around the center of the field of view.  It is advisable to center the object stage with the objective of highest magnification on the nosepiece.  When changing to a lower magnification, the deviations from perfect centration are then generally so small that they can be tolerated.

When attachment cameras are used for photomicrography, it is also possible to achieve the orientation of the image within the film frame by rotating the camera assembly.  Users of this type of camera occasionally use this procedure when their mi-croscope is equipped with a square stage.

There is one illumination method where a rotating stage (or the equivalent*) is essential and where perfect coincidence of its

______________________________

*Simultaneously rotating polars (through 360°) are equiva-lent to a rotating stage.  The Dick polarizing microscope, once made by Swift, was of this type; it is very useful for hot stage microscopy when the stage cannot very well be rotated and polars are desired.

axis of rotation with the optical axis of the objective must be maintained with the highest degree of precision: microscopy in polarized light.  Since polarized light is also used in biological microscopy, <u>i.e.,</u> for detection and optical analysis of anisotropic crystalline inclusions in biological objects or for detection of birefringence (<u>e.g.</u>, in muscle fibre), brief explanations of the special optical requirements for these applications of polarized light will be of practical value.

Birefringence is usually observed in polarized light with the polarizer and analyzer in "crossed" orientation.[*]  By rotating the object stage, the orientation of the planes of vibration of anisotropic objects with respect to the planes of vibration of polarizer and analyzer change, causing the anisotropic object to appear with varying light intensity.  Very small crystalline inclusions may be moved to the center of the field of view (point of intersection of the cross hairs of the ocular) for observation of interference figures.  In such situations when an objective of higher magnification is used the orientation of the object within the field of view must be maintained with precision.  Special rotating object stages are available for microscopy in polarized light.  They rest in precision ball bearings and have <u>no centering devices</u> to avoid even the smallest degree of lost motion.

Centration of the optical axis of the objective with respect to the axis of rotation of the stage is achieved by other means.  One manufacturer offers a multiple nosepiece with facilities for centering each objective.  Single objective carriers with centering device for the objective, for interchangeable use in the dovetail slider which is part of the microscope stand, are also available.  Furthermore, "strain-free" objectives are available for microscopy in polarized light and are supplied in special mounts with individual centering devices for attachment to a regular multiple nosepiece.

---

[*]Parallel polars may be used but colors complimentary to those observed with crossed polars will be observed.  Slightly uncrossed polars may also be useful to show both isotropic as well as birefringent objects together.

Users of microscopes with square stages will be greatly disappointed when they decide to attach a polarizer and an analyzer to their equipment and discover that the missing facility for rotating the object stage for orientation of the anisotropic object greatly restricts the value of these two components.

To complete the description of available microscopes of the first category: they also have <u>interchangeable bodytubes</u> for monocular or binocular observation and for use in photomicrography (a combination of a straight monocular tube with an inclined binocular tube, a so-called trinocular head).

The optical performance characteristics of objectives, oculars and condensers are described in Chapter 6 of Volume 14. In addition to objectives for brightfield and darkfield illumination, there are others to be used with illumination methods described in other volumes of this series. For instance, there are objectives for phase contrast microscopy. These objectives are equipped with "phase rings" — ring-shaped thin layers of a semi-transparent material in their back focal planes.

Occasionally, when selecting equipment for a microscope to be used for several illumination methods, including phase contrast illumination, when a limited budget is available, the question arises: can an objective for phase contrast be used for brightfield illumination? The best way to answer this question is to make a comparison of the images formed by the two types of objectives when brightfield illumination is used. This comparison will reveal that the image formed by the phase contrast objective is inferior for brightfield illumination. The cause of this inferiority is the phase ring and the optical effect which it produces. In brightfield illumination, coherent portions of wave surfaces from individual object points pass through the back focal plane of the objective. Within the ring-shaped zone of the phase ring in the back focal plane of the objective, a reduction of the amplitude of the waves and also a retardation of the waves is produced. Diffracted waves proceed from that plane to the image plane and cause additional interferences which impair the quality of the image. It may still be acceptable for a brief informative glance at the object area within the field of view, but for continued visual observation or for photomicrography, the image quality is unacceptable.

Also in the <u>second category</u> (reflected light microscopes) there are two types of microscopes: inverted and upright. The inverted microscope is often preferred for metallurgical microscopy (metallography). Metal specimens generally have irregular shapes. When they are prepared for microscopy, one surface is ground plane, polished, and etched to reveal surface relief structures. This plane surface must be aligned in perpendicular orientation to the optical axis of the objective. When an inverted microscope is used, this surface rests on the object stage and is correctly aligned for image formation. When an upright microscope is used, the prepared surface must be accurately parallel to the surface which rests on the object stage. Special devices (mounting presses) are available for mounting the metallurgical object in correct alignment for use with an upright microscope.

Microscopes for reflected light are equipped with vertical illuminators and built-in illumination systems. There are simple models of vertical illuminators without aperture and field stop. More elaborate models have these two components and facilities for centering the field stop and for lateral displacement of the aperture stop (for oblique illumination). Since the objective is also its own condenser, the optical axes of these two optical units always coincide. The performance of the complete illumination system for reflected light is analogous to that for transmitted light. Both systems perform on the basis of the optical principles of Köhler illumination.

Some vertical illuminators have interchangeable reflector units. In addition to the reflecting glass plate (Volume 14, page 64) and the ring-shaped elliptical mirror for darkfield epi illumination (Chapter 4, J; page 67) there are reflecting units for polarized light and others with "dichroic" filters for fluorescence microscopy in reflected light. Additional information about these units will be found in other volumes of this series.

The variety of objectives for reflected light is more restricted. Some manufacturers offer only one category. These objectives are not only corrected for image formation of uncovered objects (Volume 14, page 88), but also for flatness of field. The detrimental effect of glare is reduced as much as possible, not only by antireflection coating of the surfaces of its components but also by designing the lens system so that the surfaces of the components are shaped as favorably as possible to avoid reflections. The initial magnifications of these objectives differ from those for transmitted light because special standard magnifications have been established for metallurgical microscopy and these objectives comply with the specifications. The objectives are available in mounts, only for vertical illumination or also in mounts for additional epi illumination of high obliquity (darkfield illumination with reflected light). Because of the larger diameter of the surrounding "shell" with curved mirror to guide the light to the object plane, these objectives cannot be attached to regular nosepieces. Special objective carriers for single or multiple objectives are available for them.

In the third category (which might be called universal microscopes) there are several models of microscope stands which differ from each other in regard to the degree of versatility. Those of greatest versatility are designed for interchangeable use of light sources, including those of highest intrinsic intensity (mercury, cesium iodide and xenon arc lamps). Every other component of the microscope is easily interchangeable: condensers for all illumination methods with transmitted light as well as every type of illuminator for reflected light, polarized or nonpolarized; square or rotating, circular object stages with various types of mechanical stages; single as well as multiple objective carriers for every type of objective; monocular as well as binocular and other types of tubes etc. Catalogs of manufacturers are available with detailed descriptions of these models.

C. ALIGNING THE OPTICAL SYSTEM: The following outline of the procedure to be used for aligning the optical system of the microscope for image formation of optimum quality is based on an illumination of the object with transmitted light. It is similar to the procedure for most of the other illumination methods, not only for transmitted but also for reflected light.

1. When transmitted light is used, the object should be covered with a cover slip of the correct thickness. (No. 1-1/2 cover slips with thickness 0.16-0.19 mm meet the requirements and vary within acceptable tolerances.)

Objective and ocular should be selected for image formation so as not to exceed the limit of useful magnification (unless deviations are intentional, as explained on page 79). The category of the selected objective should be such that the best image, formed by the objective, meets the requirements of flatness of field, specimen contrast etc.

For binocular observation, the binocular tube must be so adjusted that the interpupillary distance of the observer as read from the scale between the tubes is also adjusted on the individual tubes so that the same figure coincides with the index marks for each ocular. Unfortunately, all binocular microscopes do not have the individual tube scales. If alternate observation with the right eye through the correctly adjusted right tube and with the left eye through the left tube shows that when one image is in focus the other is slightly out of focus, the observer's eyes are subject to slight differences. In this case, one of the tubes should be left in the correctly adjusted position of the interpupillary distance coinciding with the index mark, whereas slight correction can be made by turning the other tube until both images are simultaneously in focus.

Parenthetically, attention is drawn to the fact that this procedure is optically not entirely correct, because only when both tubes are adjusted for the same figure on the respective index marks are the two mechanical tubelengths identical. Unequal mechanical tubelengths also cause unequal optical tubelengths and unequal magnifications. These deviations, however, are so small that they are below the limit of detectability.

To comply with every optical requirement, no matter how insignificant, the user can obtain for his binocular microscope a pair of oculars, one of which is equipped with an adjustable eye lens. After having set both tubes to the same figure on the index mark and focusing the image through the ocular with the fixed eye lens, correction of the sharpness of the image, as observed through the other tube, can be made by adjusting the eye lens of the ocular in that tube.

Leitz supplies a combination tube (inclined binocular for visual observation and straight monocular for photomicrography) in which the mechanical device for varying the interpupillary distance is coupled with a compensating device which retains the mechanical tubelength at the same value when the interpupillary distance is changed. They also supply oculars with adjustable eye lenses.  The main advantage of this tube is that parfocality of the objectives is retained and individual resetting of the tubes of the binocular body is avoided.

For alignment of the illumination system, the condenser must be used with all of those components in the lightpath which are required for formation of the image of the field stop in the object plane.  Zeiss condensers 0.9Z or 1.3Z, for instance, should be used with the top lens as well as the "auxiliary lens" in the lightpath.  An objective of medium initial magnification should be used so that when the field stop is fully open, the entire field of view is illuminated.  When an oil immersion objective of 100X is used for alignment of the illumination system, the image of the field stop, even when it is closed as far as possible, is practically as large as the field of view.  The aperture stop of the condenser should be fully open and the field stop should be closed.  The image of the object should be focused with the coarse and fine adjustments of the microscope.  The illumination throughout the field of view will be unsatisfactory.  This image of the field stop will be out of focus and uncentered.  This is shown in Figure 42.

2.  The condenser should be moved up or down until the image of the field stop is as sharp as possible.  Its image will not be as sharp as that of the object.  Furthermore, when the image of the field stop is still slightly out of focus, color fringes are visible at its edge.  When the condenser is too high, the color of the fringe is yellowish, when it is too low, the fringe is bluish.  These fringes are caused by incomplete correction of condensers (of low price range) for chromatic aberration.  When the aplanatic-achromatic condenser is used (if its NA is higher than 1.00, oil must be provided between condenser and object slide), or color fringes will be noticeable.  When the field stop image is in best focus, but uncentered, the image will look as shown in Figure 43.

3.  The centering screws of the substage should now be used to center the image of the field stop within the field of view.  The result is shown in Figure 44.

4.  With the image of the object and the centered field stop in focus the field stop should be opened only so much that its image disappears at the edge of the field of view, <u>but not farther.</u>

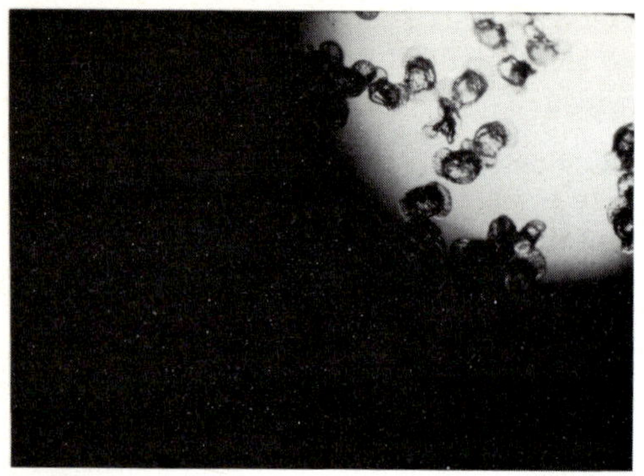

Figure 42. The field of view with an uncentered, unfocused field stop.

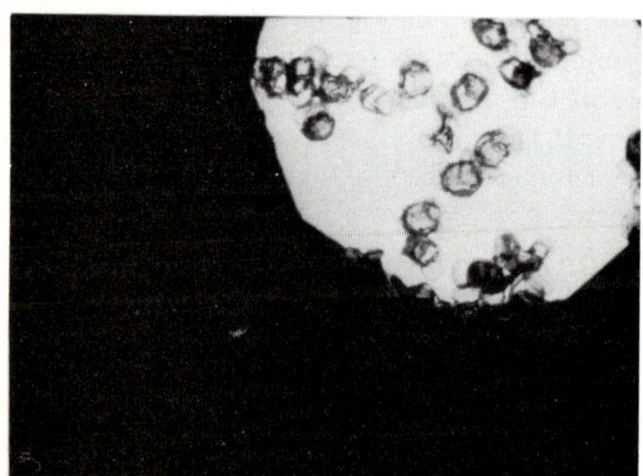

Figure 43. The field stop is still not centered but has been focused using the substage condenser.

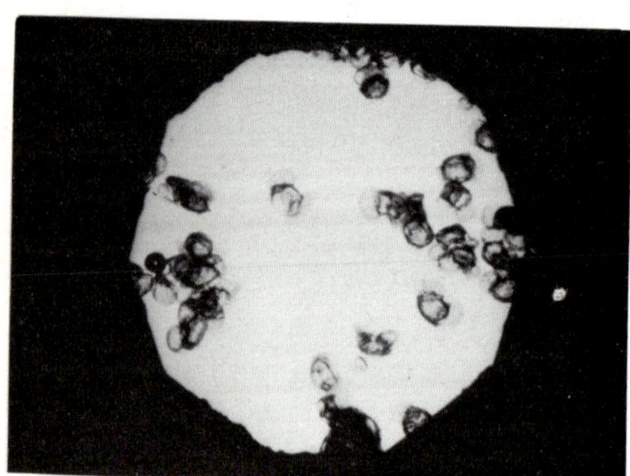

Figure 44. The field stop is now both centered and focused.

When the diameter of the illuminated area is larger than the field of view, there is no increase of the intensity _within_ this field, but the detrimental effect of glare increases.

If the objective, required for the specific task at hand, differs from that with which the alignment of the illumination system was made, it should now be swung into the lightpath. Minor correction of the centering of the image of the field stop may become necessary but the acceptable tolerances for deviation from correct centration are not very critical, because the possible increase of glare is not very great.

The condenser should always remain in the position as adjusted when objectives are changed, except for correction of centration when necessary. Any displacement of the condenser in the direction of the optical axis, _e.g._, for varying the light intensity within the field of view, _must be avoided_. Many users of microscopes adhere to this incorrect procedure. Variations of the light intensity should be made by interposing a neutral density filter or, sometimes, by varying the voltage of the light source (within certain limits and when variations in color balance are not important).

5. When all of the adjustments as described have been completed, and the aperture stop is still fully open, the image will appear as shown in Figure 45. The aperture stop should now be closed very gradually while the observer critically evaluates the image quality to detect slight improvements. The keen observer will notice that at a somewhat reduced NA of the illumination, a fairly abrupt change of the optical quality of the image occurs (see Chapter 4). The contrast between the smallest resolved object detail increases slightly, the colors of stained objects appear slightly more "saturated"; there is a slight reduction of glare; the flatness of the field is slightly improved and there is also a slight increase of the depth of field. At that position of the aperture stop, the image quality is at an optimum and the microscope performs to the limit of its capacity of the selected optical components.

The observer may close the aperture stop still further for training purposes to observe the progressive deterioration of image quality and decrease in resolving power. A certain amount of experimentation is necessary to find that diameter of the aperture stop at which optimum image quality prevails. If objectives of different categories (Neofluars, Planapochromats) of equal magnifications are available, it is advisable to use these objectives to show how the state of correction of the objective influences the diameter of the aperture stop at optimum image quality. The same procedure may also be repeated with other objects, possibly mounted in media of different refractive indices.

At lowest magnifications, the components of the condenser which are in use may not make it possible to form an image of the field stop. For instance, when Zeiss condensers 1.3Z or 0.9Z are used, the top lens as well as the auxiliary lens must be swung out of the lightpath. Under these conditions, the field stop acts as an aperture stop (Volume 14, pages 60-61). The field stop should now be used for reduction of the NA of the illumination and improvement of the image quality. Many microscopes, _e.g._, Bausch and Lomb, Olympus and American Optical Company have a special auxiliary lens underneath the condenser which should be swung _into_ the lightpath at _lowest_ magnifications. Other manufacturers may use other combinations of the condenser system at lowest magnifications.

The photomicrograph of Figure 46 shows the same area of the object as that of Figure 45, but with the aperture stop slightly closed for optimum image quality.

The time required for making adjustments of steps 1-5 is much less than that required for reading these directions. The user of the microscope who follows the described outline will be able to produce results of optimum quality with minimum effort and with certainty of success.

When a _darkfield condenser_ is used, its optical system must be focused and centered so that the oblique cones of light of all azimuths intersect the object plane in the center of the field of view.

A low power objective (10X or lower) should be used for this alignment. The immersion type condenser should be lowered with the rack and pinion motion of the substage, until the upper plane surface is about 1/4 in. below the surface of the object stage. A drop of immersion oil should be deposited on

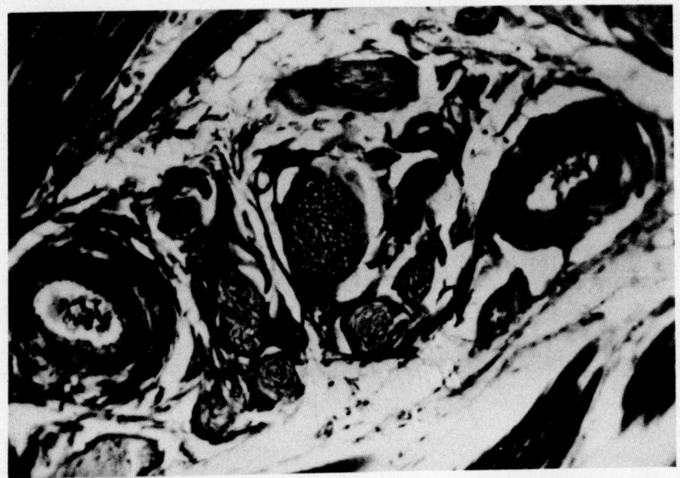

Figure 45. Aperture stop fully open. Here the field stop is centered, focused and opened to fully illuminate the field of view.

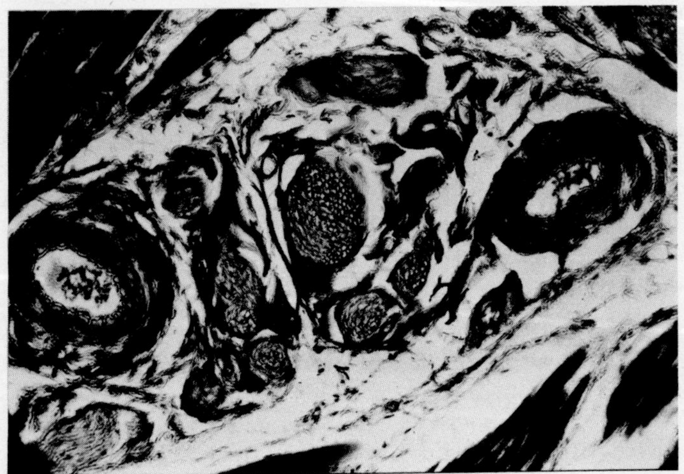

Figure 46. The aperture stop is now slightly closed for best image quality.

the top surface. The object should be placed on the object stage and the condenser should be moved upward gradually until the oil has made contact with the lower surface of the object slide. The image of the object should now be focused with the coarse and fine adjustments of the microscope. When the condenser is too low or too high, the plane of intersection of the oblique cones of the illumination does not coincide with the object plane. A ring-shaped area of the object is illuminated and the center of the ring is dark. If the illuminated ring-shaped area is not centered within the field of view, correction must be made with the centering screws for the substage condenser. By gently raising or lowering the condenser, it can be adjusted to a position where the ring-shaped area decreases in diameter until a bright, circular area of small diameter is illuminated. Figure 47 shows the ring-shaped area when the darkfield condenser is centered but not in focus. Figure 48 shows the bright spot in the center of the field of view when the condenser is centered and in focus. Another drop of oil should now be deposited on the upper surface of the object slide and the oil immersion objective with iris diaphram should be swung into the lightpath and the image of the object should be focused. The iris diaphram of the objective should be adjusted for optimum contrast and resolving power of the image (Figure 49). This model of the darkfield condenser can be used also with dry objectives of high magnifications (40X and higher). At lower magnifications, the entire field of view will not be illuminated, especially when widefield oculars are used. As previously mentioned, dry darkfield condensers are available for these lower magnifications. The procedure for aligning these darkfield condensers is similar to that of the immersion type condenser but, of course, no oil should be used between condenser and object slide or between slide and objective.

When <u>vertical illuminators</u> with aperture and field stops are used, the alignment of the illumination system is similar to that for brightfield illumination by transmitted light. Since the objective also acts as a condenser, the two optical axes always coincide, but centering devices for the field stop are available.

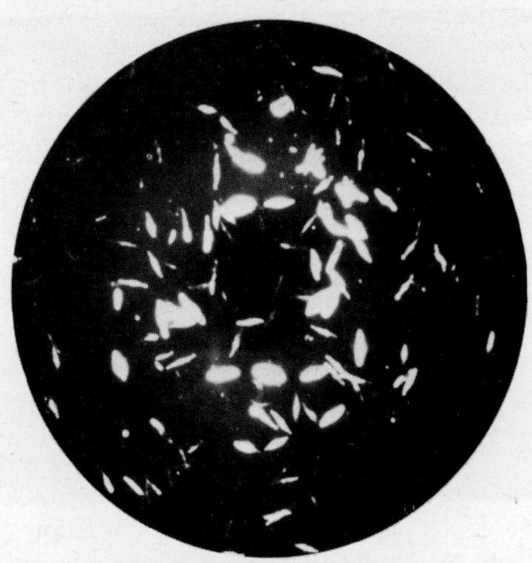

Figure 47. The darkfield condenser is not focused (Photomicrograph taken with objective of low magnification)

Figure 48. The darkfield condenser is now focused and centered.

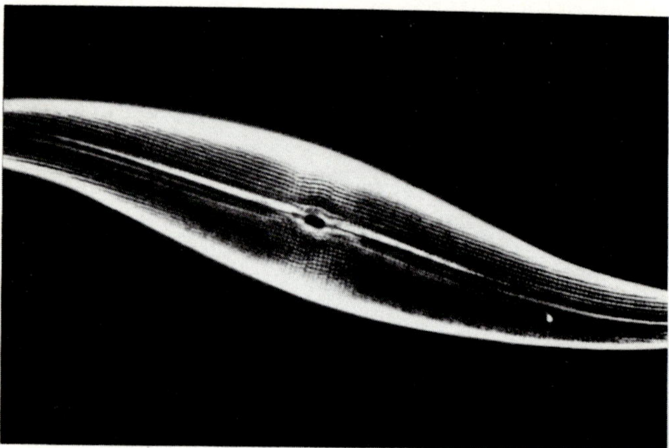

Figure 49. Changing now to oil immersion darkfield the object in Figures 47-48 now appears as shown here.

CONCLUSION: The <u>geometrical</u> relationships between the components of the microscope — from the light source to the plane of the final image (as described in Volume 14) prevail for all types of illumination: brightfield and darkfield illumination by transmitted light and for vertical as well as highly oblique illumination by reflected light.

The <u>resolving power</u> of the microscope always depends on the wavelength of the light and the NA's of objective and illumination.

The <u>optical character</u> of the image varies, depending upon the selected illumination conditions. A careful study of the illumination methods described in other volumes of this series is recommended. The most favorable illumination method should be selected on the basis of an analysis of the optical characteristics of the object. When this selection has been made, and the components of the optical system have been selected on the basis of the physical optical aspects of image formation, described in this volume, the procedure of aligning the illumination system as outlined in this chapter should be followed with modifications which may be required for these other illumination methods. The microscope will then perform to the limit of its capacity.

## INDEX

Abbe controversy, 58
Abbe diffraction
   apparatus, $\underline{x}$, 37, 42, 43
Abbe theory, 35
Adjustable transformers
   proper use, 91
Afocal illumination, 53
Aligning the optical system
   transmitted light, 97
   reflected light, 104
Amphipleura pellucida, 40
Amplitude, 3
Angstrom, 12
Antiflection coating, 64
Aperture diaphram, 33
   adjustment, 101
Artificial images, 46, 48
Axial illumination, 39
Axial resolving power, 29
Azimuth of vibration, 9
Back focal plane,
   objective, $\underline{x}$, 36
Berek, M., $\underline{v}$., 30
Binocular eyepieces, 98
Biological objects in polar-
   ized light, 94
Birefringence, 94
Blooming (antireflection
   coating), 64
Body tubes, interchange-
   able, 95
Brightfield condenser, 92
Coated glass surface, 66
Coherence, 8, 33, 55
Coherence in object
   planes, 33
Coherent light, 41
Color temperature, 92

Condenser, 33
Contrast, 64
Convex lens, 20
Cover slips, 97
Critical illumination, 51, 58
Darkfield illumination, 58,
   102
   reflected light, 72
   visibility limit, 60
Depth of field, 29, 32
Dick model, Swift
   microscope, 93
Diffraction
   Abbe test plate, 43
   coarse rulings, 44
   disc, 23, 55
   fine rulings, 44
   limit of resolution, 17
   multiple slits, 4
   one slit, 4
   patterns, $\underline{x}$, 42
   two slits, 4, 7, 11
Dispersion, 13
Dispersion staining, 63
Epi illumination, 67
Epi illumination, dark-
   field dichroic filters, 96
Empty magnification, 75
Field diaphram,
   adjustment, 100
Field of view, size, 79
First order maxima, 7
Fluorescence micro-
   scopy, 63, 96
Fourier plane, see back
   focal plane, objective
Frequency of light, 19
Frontispiece, $\underline{x}$

## CREDITS

| | | |
|---|---|---|
| Composition | — | Cleopatra Brown |
| Typing | — | Cleopatra Brown |
| Drafting | — | James Graft |
| Frontispiece | — | Walter McCrone |
| Proofreading | — | Lucy McCrone |
| Editorial Assistant | — | Sylvia Graft |
| Printing | — | Newnorth Artwork Ltd Bedford |